THE LUSHAI HILLS
Annexation, Resistance and Pacification 1886-1898

LOCATION OF LUSHAI HILLS

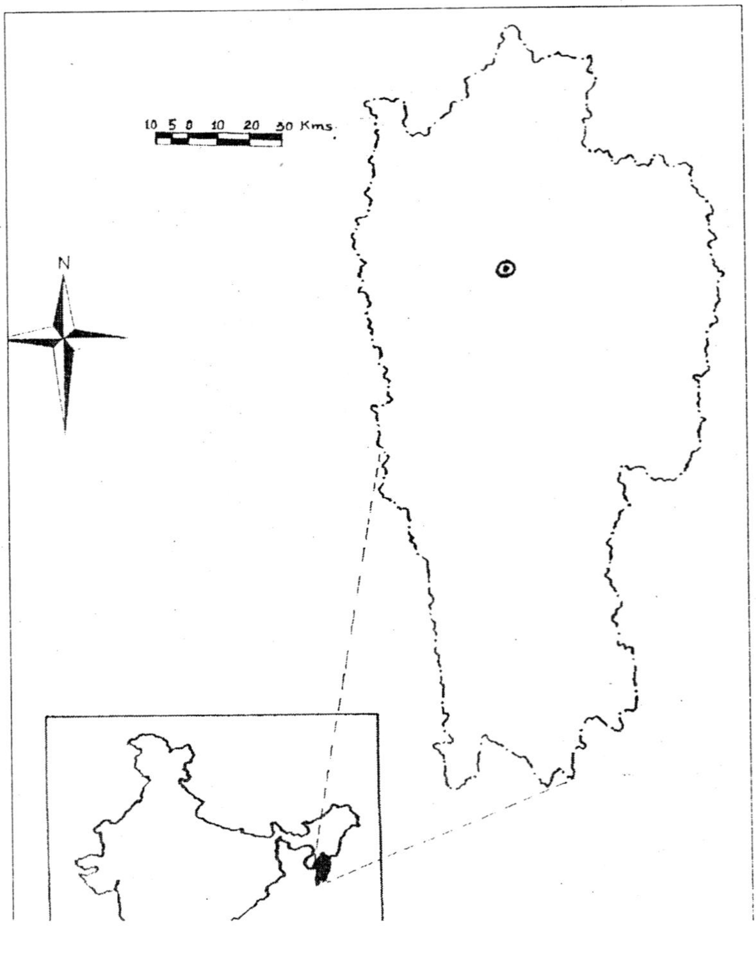

THE LUSHAI HILLS
Annexation, Resistance and Pacification 1886-1898

C. LALTHLENGLIANA

AKANSHA PUBLISHING HOUSE
NEW DELHI-110002

AKANSHA PUBLISHING HOUSE
4649-B/21, Ansari Road
Darya Ganj, New Delhi - 110 002
Ph.: 23263193/9811582579
Fax: 011-23263193
Email: ektabooks@yahoo.com

The Lushai Hills, Annexation, Resistance and Pacification 1886-1898

First Published 2007

Rprint 2025

ISBN 978-81-8370-111-2

PRINTED IN INDIA

Published by M.P. Misra for Akansha Publishing House, New Delhi and Printed at Aryan Enterprises, Delhi

TO MY PARENTS
C KAPHLEIA AND CHALHMINGTHANGI

PREFACE

Most recent publications on the history of British rule in the hill areas of North East India have virtually ignored the fact that almost everywhere the hill people took up arms to resist colonial rule. Sir Alexander Mackenzie's *History of the Relation of the Government With the Hill Tribes of North East Frontier of Bengal* and Sir Robert Reid's *History of the Frontier Areas Bordering on Assam* gave little or no importance to the subject. Nor does Professor H.K. Barpujari's three excellent volumes on British policy in the North East Frontier devote much space to tribal resistance. Works specifically on the Lushais such as Suhas Chatterjee's *Mizoram Under the British Rule* and BB Goswami's *Mizo Unrest* are equally silent on the nature of opposition to British rule. Yet tribal resistance to imposition of colonial rule was at places intense and at others protracted. Thus with the exception of a recent study of the Jaintias, by Shoban N. Lamare *The Jaintia Resistance of 1860-1863*, the subject remains so far unexplored.

The present work attempts at a critical analysis of the nature and extent of Lushai resistance to the imposition of colonial rule. Resistance here is seen as opposition before the essential apparatus of colonial administration is fully introduced and takes shape; as opposed to rebellion which is action at a later date when administration is already in place. The confusing terms primary and secondary resistance are avoided. The work opens with an account of the frontier disturbances which ultimately led to the occupation of Lushai

Hills. It further analyses the beginning of the British Administration and the uprising of the Western Lushais which followed. It tries to show how demands by the British provoked the uprising. The result of the extension of the system to the Eastern Lushais is examined in the next. Particular attention has been paid to the methods applied by Robert McCabe to crush the Lushais. The Pawi and Haulawng uprising too is another interesting chapter that follows. That fines and impositions were once again at the root of the troubles with the Lushais is to be seen in case of the uprising of the Kairuma group. The importance of disarmament in the pacification process is also examined in the chapter. The beginnings of British administration and the impact on it of the experience of Lushai resistance is shown in the final chapter. The Introduction provides the necessary background to the study, whereas the Epilogue summarises the findings. The title needs some explanation. The term 'Lushai' actully refers only to one tribe, the Lusei; the generic term of the tribal people living in the state of Mizoram today being 'Mizo'. However, Lushai and Lushai Hills have been used only because of their currency during the period under review and appears in all documents quoted and the use of two different names for the same people and region would create confusion.

This work is based primarily on documentary sources preserved in the National Archives of India, New Delhi, Mizoram State Archives, Aizawl, District Records Office at Lunglei, and Secretariat Records of Assam. Extensive use has also been made of memoirs, journals, and contemporary writings. A select bibliography is appended at the end of the volume.

This work is based on my doctoral dissertation submitted to North-Eastern Hill University, Shillong. It is difficult to express my feelings of gratitude to those who had rendered invaluable help towards the completion of the work. Throughout the preparation of this book, Professor Imdad

Hussain ceaselessly gave me learned guidance, constant inspiration, editing and encouraging motivation. I am deeply beholden to him. I am immensely grateful to my wife, and father, for their constant inspiration, particularly at times when I found my task *daerating* and despised of completion. I am also grateful to my friends Dr J Zorema, Dr S R Zonunthara, Dr Sangkima, Mr. R Lalzawnga, Mr. Lalnghenga, and Dr. Lalngurliana Sailo, who had lent me a great help in finding the source material and giving me encouragement with inspiring words and without whose help this work may not be materialised. Finally, I am grateful to Mr. M P Misra, Akansha Publishing House, for agreeing to publish the book and bringing it out in record time.

Aizawl, Mizoram. C. LALTHLENGLIANA

CONTENTS

ABBREVIATIONS

FPAP	—	Foreign Department Political 'A' Proceedings
FEAP	—	Foreign Department External 'A' Proceedings
FSEP	—	Foreign Department Secret 'E' Proceedings
FPEP	—	Foreign Department Political 'E' Proceedings
FEBP	—	Foreign Department External 'B' Proceedings
MSR. Pol.	—	Mizoram Secretariat Record Political Branch
MSR. G.	—	Mizoram Secretariat Record General Branch
ASR. Foreign ment	—	Assam Secretariat Record Foreign Depart-
D.C. 'A's	—	Deputy Commissioner's Office Record, Assam
KW(s)	—	Keep With(s)

LUSHAI HILLS
(1890-1898)

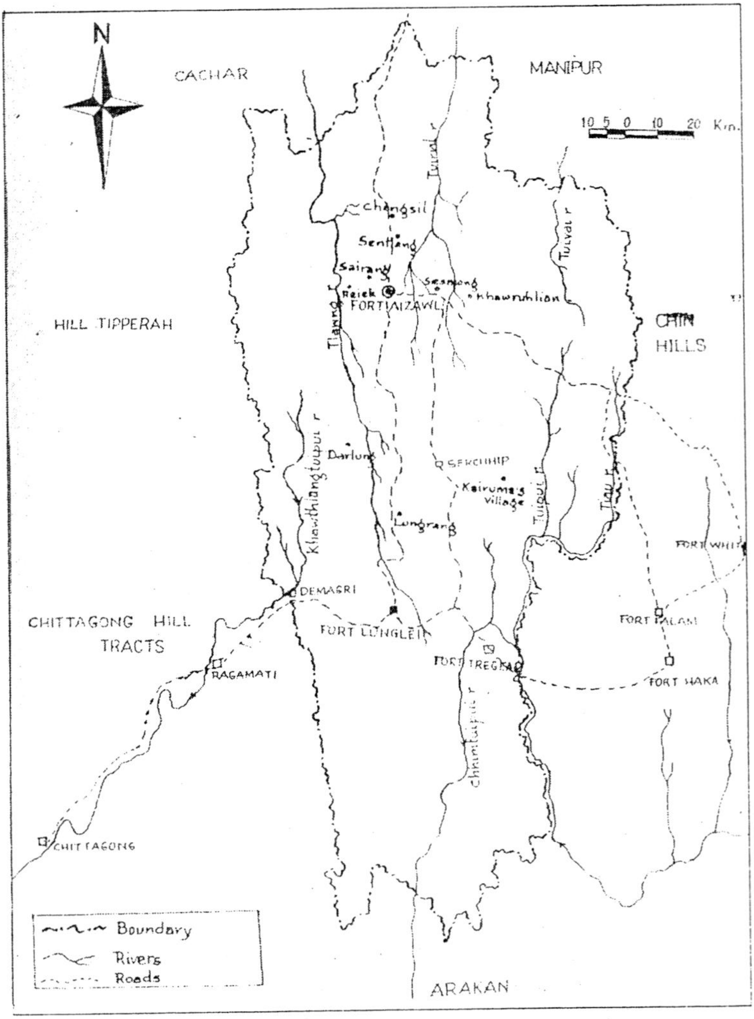

INTRODUCTION

The Lushai Hills, later Mizo Hills district of Assam and now the state of Mizoram, lies in the extreme southern corner of North East India. It covers today, an area of 21,081 square kilometres. It lies between 21° 56' - 24° 31' north latitude and 92°16' - 93°26' east longitude.[1] The tropic of Cancer runs about 50 kilometres south of Aizawl, the capital city of the hills. It is bounded by Cachar district of Assam on the north, by Manipur on the north-east, and by Tripura on the north-west. It is sandwiched between Myanmar on the east and south-east and Bangladesh on the west and south-west. About seventy five percent of its 573 kilometres long boundaries are international.

The hills consist of small ranges running in the north-south direction in parallel series separated from one another by narrow valleys. The entire area with the exception of a small portion lying in the plains are steep hilly terrain. The ridges of the ranges of hills vary from about 40 to 2000 metres in height. The Blue Mountain, called by the people, 'Phawngpui,' situated in the South Lushai Hills, is the highest point at 2157 metres above sea level.[2] The valleys are narrow, rugged, and unsuitable for wet rice cultivation. Excluding the

---

[1] *Mizoram Statistical Handbook*, 1974 , (Aizawl, 1974) p.3. see also Rintluanga Pachuau, *Geography of Mizoram*, (Aizawl, 1994), p. 23.

[2] Rintluanga, *Ibid*.

small portion of the plain areas in the extreme north, the whole area comprises of small hills and narrow valleys. Within the hills the only few small plain areas are mostly located in the eastern region. Champhai valley, 11.27 kilometres in length and 4.83 kilometres in width is one such plain area. Another is Vanlaiphai situated in the south eastern corner. Thenzawl is the third important level area in the middle of the region.[3]

The rivers are small and almost entirely rain-fed. Hence, a number of them dry up in the winter season. As such water communication is almost impossible during the winter. However, a few rivers such as Tlawng, Tuirial, Chhimtuipui and Khawthlangtuipui[*] are navigable for small crafts throughout the year.[4] But in the rainy season the rivers swell up rapidly making them unfordable. Communications become extremely difficult then.

The Lushai Hills, located in the tropics, enjoyed a fairly pleasant climate. The range of temperature between summer and winter was only slight. The monsoons, which generally prevailed during the months of May, June and July made the summer less hot. The hills enjoyed abundant rainfall. The average recorded in the northern hills was 80 inches in a year, whereas the southern received heavier rainfall and 131 inches annually was usually recorded.

Climatic conditions of these hills, being mild and humid, accounted for the dense forests of evergreen trees and bamboo. Therefore, forest played a vital role for their daily livelihood. The abundant growth of bamboo provided considerable household materials of the inhabitants. Indeed, the life of the inhabitants were largely governed by the bamboo forests. Colonel Thomas Lewin, who for many years was

---

[3] *Ibid.*

[*] Non-Lushai name for these rivers are *Dhaleswari* for Tlawng , *Sonai* for Tuirial, *Kolodyne* for Chhimtuipui and *Karnafuli* for Khawthlangtuipui.

[4] *Ibid.*

Superintendent of Chittagong Hill Tracts and knew the Lushai Hills, thus writes:

> The bamboo is literally his staff of life. He builds his house of the bamboo; he fertilises his field with its ashes; of its stems he makes vessels in which to carry water, with two bits bamboo he can produce fire; its young and succulents shoots provide a dainty dinner dish; and he weaves his sleeping mat or finds slips thereof. The instruments with which his women weave their cotton are of bamboo. He makes drinking water cup of it, and his head at night rests on a bamboo pillow; his forts are built of it, he crèches fish, makes baskets and stools, and thatches his house with the help of the bamboo. He smokes from a pipe of bamboo; and from bamboo ashes he obtains potash.[5]

In the lower altitudes the forest was mainly thick groves strewn with wild plantains, whereas dense wooden forests were to be found in its higher region. Among the principal species of trees found in these forests was rubber which became an important item of trade. A number of species of bamboo and orchids of various kind, *Bauhinia variegata*, *rhododendrons*, yellow sunflowers and many other colourful wild flowers were to be seen in abundance.[6]

The Lushai Hills was the home of a variety of wild life: elephants, bisons, tigers, leopards, bears, various kinds of deer, wild boars, were found in its jungle.[7] Birds of various kinds which included rare species like the hornbills, pea-pheasants, sunbirds, and the like, were also to be found in the jungles.

---

[5] T H Lewin, *Wild Races of South-Eastern India*, (London, 1870). pp. 15-16.

[6] Rintluanga, *op.cit.*, pp.52-53. One peculiar features of the bamboo forest was that at intervals of about fifty years, they flowered, which led to the multiplication of rats which fed on the flowers, and in turn caused famine as they devoured all crops.

[7] *Ibid*.

PEOPLE

The Lushai Hills, was inhabited by a number of tribes and sub-tribes speaking different dialects of the same language. The major tribes were the Luseis, Pawis, Lakhers, Hmars, Paite, Thangluah, Palian and Zadeng.[8] All these were actually known by the generic name "Mizo."

For sometime before the British came in contact with these hills, there was a northerly movement of tribes. Colonel Frederick Lister, the Political Agent of the Khasi Hills who led an expedition into the Lushai hills during 1849-50 thus reported:

> It would appear that the tribes to the south have been gradually driving one another in a northerly direction; for, first, some Nagas that were located in the Boobun hills and in southern Cachar were obliged by the Tangune Kookies to flit and to take up their abode in the hills north of the Borak, when the Tangunes took possession of their ground, and they having in their turn been driven up by the Chansen and Tadoe tribes, the Tangunes were also afterwards obliged to vacate and to move on into the northern hills.[9]

Lister goes on to say that after them the "Chansens" were obliged to do so likewise; and the "Tadoes," who had been driven up by the "Luchyes," a very powerful tribe, first settled about seven years since within eight and ten miles south of Silchar station, and became Company's riots, and made themselves useful by cutting timber, bamboos, cane, and others, which they used to bring to market. Lister further reported:

---

[8] The earliest notice of these tribes is to be found in John Rawlins, "On the Manners, Religion and Laws of the Cucis or Mountaineers of Tipra," *Asiatic Researches*, Calcutta, 1972, Vol. II, No. XII; John Macrae, "Account of the Kookies or Lunctas," *Asiatic Researches*. Calcutta, 1801, Vol. VII. No. VI; See also C.A.Soppit, *A Short Account of the Kuki Lushai Tribes*. (c. 1893).

[9] Quoted in Alexander Mackenzie, *History of the Relations of the Government With the Hill Tribes of the North East Frontier of Bengal*, (Calcutta, 1884), p. 287.

> but after having been located there for some four years, the "Luchye Kookies" in November 1849 attacked them, burnt three of their villages, killed several of the inhabitance, and took away several of them in to slavery, and then the whole of the "Tadoe" tribe flitted, left the south and settled down in the northern hills.[10]

Since the British first came into contact with the various sub-tribes of Kookies, the people living in these hills were referred to in the early years as "Kookies". It was Lister's expedition which first established the existence of the Lushais or more correctly the Lusei, as a separate and distinct tribe. From then and until the Expedition of 1871-72 the British both in Cachar and in the Chittagong Hill Tracts dealt only with the Lushais who under the chiefs had firmly established themselves in the hills. Hence the term Lushai gained currently and their land was referred to as the Lushai hills. Major John Shakespear, of whom more later on, thus points out:

> The term Lushai as we now understand it, covers a great many clans, it is the result of the incorrect transliteration of the word Lushai, which is the name of the clan, which, under various chiefs of Thangur family, came into prominence in the eighteen century...Among the people themselves the Lusheis are sometimes spoken as Duhlian...and the general population of the hills is spoken as Mizo.[11]

From the eighteen seventies the British became aware of that the people in the southern hills were different. They were then called Shendus, a term which seemed to have covered all the Pawis and not only the Lakhers.[12]

---

[10] *Ibid.*

[11] Lt. Colonel J. Shakespear, *The Lushei Kuki Clan*, (London , 1912, reprint, Aizawl , 1975). See Introduction. However as long ago as in 1892, Robert McCabe, Political Officer, North Lushai Hills in a report wrote that "The Lushais call themselves Mizo or Mizau, which term includes the following castes: Ralte, Molbem, Khuangli, Paithe, Taute, Jahau (Yahow) Dulien, Lakher, Fanai (Muallianpui), Poi, Dalang, Tangur, Sukte, Mar, Falam (Tashon), Paukhup, and Lielul. For detail see FEAP. December 1892; Nos. 42-46.

[12] N.E.Parry, *The Lakhers*, ((Shillong , 1932,), pp. 5-6,

Ethnologically, all these tribes belong to the Mongoloid stock[13] speaking the Tibeto-Burman family of language.[14] Lusei or Duhlian, which is the dialect of the Lusei clan, is now the lingua franca and is understood in many parts of the adjoining districts.[15] However, the Lusei language has been considerably modified as a result of interaction with other groups. The Ralte and number of other clans began to lose their identity in terms of the dialect and became Lusei speaking groups. Few clans, such as Hmar, Pawi, Ralte, Lakher and the Paihte still maintain their own dialects in the remote areas. However, under colonial period, the Lushais dominated. Their numerical superiority and better economic status, accounted for this, a position they maintained in the postcolonial period as well. The term Lushai therefore is applied, in this dissertation to the Lusei as well as to the Pawis and Lakhers in the south.

THE CHIEFS

The Lushais were not governed by any central authority but each village was ruled by a chief. The Sailo chiefs were the most numerous and had emerged as the most dominating ruling clan long before the British entered the Lushai hills.[16] Others like Thangluah, Palian, and Zadeng chiefs ruled over lesser number of villages, whereas the Pawis and Lakhers maintained their rule in the south and south east Lushai hills. The age-old traditional and customary laws largely governed the Lushai chiefs and his village administration. Shakespear thus wrote:

> The chief was, in theory at least, a despot, but the nomadic instinct of the people is so strong that any chief whose rule was unduly harsh soon found his subjects leaving him, and he was therefore constrained to govern according to custom.[17]

---

13 Rintluanga *op.cit.*, p.

14 G.A.Grierson, *Linguistic Survey of India, Vol. III. Part III.* (Reprint, Delhi 1976), p.3

15 J.Shakespear, *op.cit.*, p.112.

16 N.E.Parry, *op.cit.*, pp.248-249.

17 J.Shakespear, *op.cit.*, p.43.

However, the chief could not misuse his power if he desired to retain his position for custom provided that a villager could transfer his allegiance to another chief at will if dissatisfied with his chief. This is called "*peming.*"[18] Hence a chief's village became large or small according to his relations with his villagers.

Every chief ruled his village with the help of elders known as the '*Upas,*' their number varying from village to village depending on their size. They were popularly called by the British as '*mantr*i' while their counterparts in the south were referred to as "*Karbaris*". These *Upas* were appointed by the chiefs and were liable to be removed by him. Below them were other village functionaries.[19]

The relations between chiefs and his people were governed by well defined customs. A chief was entitled to certain dues in kind: '*Fathang*' a basket of grain due to be paid by the villagers and '*Sachhiah*' a portion of meat due,[20] and subsequently, others like '*Chi chhiah*' a share of salt extracted from the salty water, '*Khuai Chhiah*' a share of bee hives or honey, and the like became a sort of contribution existed among the Lushais.[21] Similarly with regard to the construction of his house, the chief was entitled to call upon the villagers

---

[18] *Ibid.*

[19] *Ibid.*, The chief and his officials therefore formed the privilege class of the Lushai society. The first and foremost groups of the privilege class, with the exception of the chief, were the *Upas*. They were exempted from the payment of the paddy due to the chief and enjoyed priority in selecting *jhum* fields. The '*Tlangau*' or the village crier was a privileged person. His duty was to proclaim the chiefs' order to the people by crying all around the village. He was entitled to receive a small amount of paddy from each house. Another was '*Thirdeng*' or the blacksmith, who was to repair and shape up all the working tools of the villagers. For such work he received a basket of paddy from each household per annum. *Thirdeng* also received a small share in every animal killed by the villagers. The '*Puithiam*' or village Priest, performed sacrificial ceremonies and offerings for the villagers. '*Ramhual*' who advised the chiefs where *jhum* should be cut each year were other important village functionaries.

[20] *Ibid.*, p.83.

[21] N.E.Parry, *op.cit.*, pp. 248-249.

to do the work without payment. The work, however, remained voluntary as no chief could force his subjects to do the work. Beyond what was permitted by custom no chiefs had the power to levy any tax or impose any other obligations.[22] In deed life among the Lushais was governed by a code of conduct known as "*Tlawmngaihna*," this was voluntary services to individuals and to the society without expectation to receive any reward in return.[23]

The main occupation of the people was agriculture and they followed shifting cultivation or *jhumming*. Besides rice, other crops that Lushai *jhum* produced were maize, cotton, various fruits and eatable roots. In no village were there any shop-keeper, and, with the exception of the blacksmith, no craftsmen. Fresh meat was obtained by hunting.[24] There was hardly any trade except but at the border *hats* or markets forest and hill produce such as rubber was bartered for such items as salt. These *hats* were maintained by people from the plains.

By the 1880's the Lushai Hills was roughly grouped under three distinct geographical areas: the Western Lushais, the Eastern Lushais and the Southern Lushais. The western Lushais were mainly the descendants of Suakpuilala[*] and their dependant chiefs. Their country covered the areas between Tuirial river on the east; district of Cachar in Assam on the north; Hills Tipperah and Chittagong Hill Tracts on the west and an imaginary line east-west through Darlung peak, about the middle of Lushai Hills in the south. The Eastern Lushais were the descendants of three chiefs namely, Vanhnuailiana, Lalphunga and Vuta, all of them from the same clan. However, among the eastern Lushai chiefs, the descendants of the first two chiefs formed a separate group

[22] L.B.Thanga, *op.cit.*, p.11.

[23] N.E.Parry, *op.cit.*, pp 19-21.

[24] J.Shakespear, *op.cit.*, p. 16.

[*] Suakpuilala was the undisputed leader of the Western Lushai chiefs during the first half of the nineteenth century.

and occupied the northern part of the Eastern Lushai Hills, whereas, the sons and grandsons of Vuta formed another group and settled in the southern part of the eastern Lushai Hills. As a whole, the eastern Lushai Hills roughly covered the areas between Tiau river, which flows between Lushai hills and Burma, in the east; the western Lushai country, using Tuirial river as boundary on the west, the Cachar district and Manipur on the north and north-east; and on the south the same imaginary line, with that of the western Lushai southern boundary.

The Southern Lushai Hills contained mixed tribes. On its extreme north and north west was the Haulawng group and their dependants who were the descendants of Tlutpawrha, a Sailo chief. Immediately east and south east was inhabited by the Pawis or Fanais known to the British as "Muallianpuis." The Lakhers or Mara known by the Arakanis as "Shendus" settled in the extreme southern corner, south of Haka sub-division and in the extreme north of the Arakan Hill Tracts.[25] However, as earlier pointed both the Pawis and Lakhers were confusedly clubbed together as Shendus as a whole the Southern Lushai Hills was bounded by the Chin hills on the east and South-east, the Chittagong Hill Tract on the west and south-west and the Arakan Hill Tracts in the south. Its northern boundaries, however, was that of the southern boundaries of the North Lushai Hills.

WARFARE

The frequent movement of tribe, both because of their migratory nature as well as because of the exhaustion of *jhum* lands or competition for scarce resources had resulted in frequent inter village or inter-tribal warfare. Village sites were therefore carefully chosen with a view to easy defence. Thus the Lushais built their villages on top of ridges while the Lakhers of the southern Lushais constructed theirs on

[25] N.E.Parry, *op.cit.*, p. 1.

the slopes of the high hills.[26] Lakher villages are however more permanent, unlike the Lusei's as Neville Parry points in his study of the Lakhers:

> The villages are permanent and are rarely moved as the Lakhers are attached to their village sites and dislike abandoning the graves of their ancestors. While the Lushei moves his village to a new sites as soon as he has worked out the surrounding land, the Lakher prefers to keep a permanent village and to spend the greater part of the cultivating season in a hut built in his field.[27]

Consequently villages in the south were very large consisting of more than 3000 houses, Selesih village was said to have been constituted of 7000 houses.[28]

These villages were elaborately stockaded, the fortifications usually encircling a village, except where the ground made it unnecessary.[29] The stockades were constructed of stout timber driven into the ground and tied with cane. The ends of each post was sharpened to a point. These were high enough to dissuade escalading. Bamboo defences, though temporary affairs, were used to reinforce the timber.[30] Parry thus describes Lakher system of fortifications:

> Every village had its fort or "*Ku*", to which the people retired on news of a raid. This fort was built in the middle of the village, and consisted of a strong stockade of tree –trunks and saplings about ten feet in height. The saplings were planted in two or three rows, so as to make the forts bullet-proof, and the stockade was loop-holed to enable the defenders to fire. All round the

---

[26] *Mizoram District Gazetteers 1989*, Directorate of Art and Culture, (Government of Mizoram), p.63.

[27] N.E.Parry, *op.cit.*, p.60.

[28] K. Zawla, *Mizo Pi Pute leh An Thlahte Chanchin*, (Aizawl , 1964), p. 21.

[29] J. Shakespear, *op.cit.*, p.57. The art of making stockade of the village was started from the Chin Hills and spread to the Lushai Hills.

[30] Imdad Hussain, "Problem of the Frontier, 1822-1860", in H.K.Barpujari, (Ed.) *The Comprehensive History of Assam, Vol. IV.* (Calcutta , 1992), p.336.

> stockade a trench was dug and sown with bamboo stakes called "*seu*," and was crossed by a draw-bridge, which was raised and lowered by cane ropes.[31]

Parry was writing of the Lakher village long after firearms had brought changes in fortifications. But the traditional weapons were spears, bows and arrows, and *daos** or long knives.[32] Shields used were made out of the hide of *gayal*, stretched in the sun to dry and then hardened and well polished. These measured more or less two feet by four feet and useful against spears and arrows.[33]

Stone chutes known as "*Sahbuak*"* were often used for defensive purposed. As the topography of the Lushai hills was favourable to its use, this was found quite effective. During the Lushai Expedition of 1871-72, for instance the expeditionary force under Lieutenant Colonel Thomas Lewin against the Lushai chief Lalzika had found to its dismay the effectiveness of this type of warfare.[34]

The first reference to firearms was around 1790. The Pawis in the Chin hills were said to have possessed guns before the Lushais: Thanghlianga, chief of Halkha, was said to posses a gun, called as "*Paranpa*," and so were much feared by the Lushai.[35] Mangpawrha, Lushai chief of Reiek, is also said to have a gun around 1812.[36] These are the earliest

---

[31] N.E.Parry. *op.cit.*, p.63. Bamboo stakes, elsewhere called *panjis*, was a common feature in North Eastern India.

* A small curved knife sharpened on one side and whose shape and size considerably varied was common to all hill tribes in the North East. But the "*Kawlhnam*" of the Lushai is said to come from Burma: "*Kawl*" means Burma and "*Hnam*," knife.

[32] *Ibid.* p. 48.

[33] Liangkhaia, , *op.cit.*, p. 27.

* For the use of stone chutes among the Abors or Adis as they are now called, see *Frontiers and Overseas Expedition From India*, Vol.VII, The Official Account of the Abor Expedition, (Reprint Delhi, 1983), facing p. 38.

[34] FPAP. August 1872; No 114-126, Dairy of Civil Officer, Right Column Lushai Field Force : Captain T.H.Lewin. 30 December 1871.

[35] F.Lalremsiama, *op.cit.*, p.47.

[36] *Ibid.*

instances of the use of firearms in these hills. But it was only after contact with the British that its use became widespread. Colonel Lister who penetrated deep into the Lushai hills during his 1850-51 expedition thus reported:

> The fighting part of the Lushai population are composed, first, of Lushais, who appeared to be a cross between the Kookies and Burmese; secondly, of a certain number of true Burmese, entertained for the purposes of warfare; and, thirdly, of refugees and outlaws from Munipore and our own frontier...The Burmese portions of the force are armed with muskets and *dows* the remainder with spears and *dows*.[37]

The muskets used by them were flint-locks, the powder being manufactured by themselves out of the local materials.[38] The method of manufacturing gunpowder was simple. They collected the saturated soil out of animals' urine and dung which were mixed with soil in a large baskets poured with water. The water, of a red colour, was boiled for two or three hours and then left to cool. The nitrate crystals dried in the sun was mixed with charcoal. The mixture was pounded in a mortar and a little water and spirit were added. When it had been sufficiently pounded and dried in the sun it became ready for use as gunpowder. The powder made by this process was said to be quite powerful, but produced out a good deal of smoke.[39]

BRITISH EXPEDITIONS : THEIR IMPACT

The British came into contact with these tribes not long after the acquisition of Chittagong in 1760.[40] Not much was known of them even after many years of occupation of Cachar (1832), except that that district and Sylhet often faced their raids on their border villages.[41] C.E.Buckland thus gives the colonial perception for the disturbances:

---

[37] Colonel F.G.Lister's report, quoted in Alexander Mackenzie, *op.cit.*, p. 293.

[38] FPAP. April 1873; No. 228-234, Baboo Obhoy Churn Mitter, to the Deputy Commissioner of Cachar.

[39] N.E.Parry, *op.cit.*, pp.134-136.

[40] For details, see Imdad Hussain, *op.cit.*, p. 150.

[41] *Ibid.*

> A private quarrel with a neighbouring clan, a scarcity of women and domestic servants, and the consequent necessity of procuring a requisite number of captives to supply the wants of the tribe, the simple desire for plunder, and of obtaining heads to grace the obsequious of some departed chieftains, were the principal causes of raids.[42]

These increased in frequency and intensity by the 1840's and in 1844, Captain George Blackwood with a party of Sylhet Light Infantry and the armed Civil Police marched into the hills and punished the raiders and captured their chief Lalsutthlaha referred to in the early literature as a "Poitoo" chief.[43] The second punitive expedition, against Ngura of Sentlang for his raid on the Thado chief Kaihmunga who had settled in British territory was under Colonel Frederick Lister, referred to earlier, in 4 January 1850.[44] This expedition brought peace to the Lushai frontier for about twelve years. The third expedition in 1869 was however, sent from Cachar in two columns: one composed principally of Police under Edward Baker, Deputy Inspector General and the other under Brigadier General Nuthall, consisting of the 44th and 7th Native Infantry and artillery. Both the parties failed to punish neither Vanpuilala nor Suakpuilala, who were suspected to have involved in a raid on the Kala Naga stockade.[45]

Throughout the 1860's various methods were tried out to secure the British frontier, but each proved of no avail. Much of the disturbance of the period was due to the extension of tea gardens deep into what the Lushais understood to be tribal territory. The result was that the tea garden became the target for Lushai raids.[46]

---

[42] C.E.Buckland, *Bengal Under the Lieutenant Governors, Vol. I*, (Calcutta, 1901) p.180

[43] R.G.Woodthrope, *op.cit.*, p. 12.

[44] *Ibid*. pp. 13-16.

[45] *Ibid*. pp.11-27.

[46] FPAP. August 1872; No. 69-78.

> Lushai raids on tea garden became particularly daring during 1870-'71. Between 22-24 January 1871 the Lushai made one daring raid on the village of Ainerkhal in Hailakandi sub-division and the factory at Alexandrapur where they murdered tea-planter Robert Winchester and carried off several captives including his little child Mary Winchester.[47]

Hence, the government began to take up the matter on a large scale, and resolved to send a larger punitive expedition known as "The Lushai Expedition of 1871-72" to cover up the whole Lushai Hills from North to South. The Northern Column was known as the Cachar Column under Brigadier General G. Bourchier; while the Southern Column called the Chittagong Column was under the Command of Brigadier General C.H. Brownlow.[48]

This expedition achieved in punishing many of the offenders of the late raids and recovered the white captine Mary Winchester from the Haulawng group of the southern Lushais. The expedition further achieved the submission of number of Lushai Chiefs under them.[49]

The Lushai war tactics was greatly revolutionised by the British expeditions. The usual Lushai method of ambush determined only throwing the whole weight to smash the enemy's potentials from certain point of their concealment at single stroke and then rapidly retreated. When the British exposed their military power to the Lushai people, it was clearly exhibited that the concentration in one front and neglecting the others would in no way proved successful but would rather become too ineffective. The British forces came in separate contingents with the detachment of the troops made successively one group after another that may support each other when a particular group was attacked on the way.[50] In such circumstances, the Lushai method of hit hard and

[47] H.K.Barpujari, *Problem of Hill Tribes North East Frontier, 1873-1962, Vol.III.* (Gauhati, 1981), p. 29.

[48] Suhas Chatterjee, *Mizoram Under the British Rule,* (Delhi, 1985) p.67.

[49] *Ibid*. p.71.

[50] FSEP June 1892; No. 80-100, No.88.

run policy may not be successfully utilised as the British forces could back up each other. Hence, it was necessary for the Lushai to adopt the British system of diplomacy. They began to take careful measure and study the movements of the *sepoys*. Accordingly, they planned the course of action to match up the British style of advance.[51]

With the contact of the British and the inter-tribal feud came to its height, the construction of stockades greatly improved. The system of double rows of stockades with earth and stones in between was introduced with the spread of firearms. The dry ditch, filled in with sharp pointed spikes of bamboo or wood, impeded initial attack. In the middle of the village, the Lushais constructed a thick wooden wall stronghold meant for the hiding place especially for women, children, and others to seek refuge when there is danger from raid. This Fort was known as "*Kulhbing.*"[52] It was always supported by the sentry posts where the security guards easily see and attack the enemies as they approached and it was backed up by the thick wooden plank. In addition to this, the whole village was encircled with the thick wooden wall, high enough that raiders may not easily climb through. This fortress was known as "*Khaw Kulh*"[53] or outer stockade. If a particular village had no enemy and the situation was peaceful, the "*Khaw Kulh*" was usually made with bamboo. About two metres surrounding the *Khaw Kulh* was cleared off all trees and pointed bamboo spikes had been placed instead.[54]

The village Fort was always associated with the sentry post known as "*Ralven Buk.*"[55] These sentry posts were not necessarily of the same design, some villages planted a thorny and thick trees whereas after the widespread of firearms, a trench were dug outside the stockades, which provided a

[51] *Ibid.*

[52] James Herbert Lorrain, *Dictionary of the Lushai Language*, (Calcutta, 1940), p. 275.

[53] Lalthanliana, *Mizo Chanchin, (Kum 1900 Hmalam)*, (Aizawl, 2000), p.201.

[54] *Ibid.*

[55] *Ibid*

protective place to repel the invaders and safeguard from the volley. These '*Ralven Buk*" were the inner part of the sentry. However, whenever they needed to retreat, an easy way to retreat inside the stockade was necessary, for that purpose a gateway was created and in fact it was a doorway or a main gate for the whole fort usually opened during day time and closed at night. This was known as "*Kulh Kawngka.*"[56] Besides the inner *Ralven Buk* the outer defence designs were still created on the branches of the trees, which were carefully watched by the young men turn by turn.[57]

By the time of the expedition of 1871-72 firearms came to be greatly appreciated. Woodthorpe thus records:

> It is only within the last fifteen years, or thereabouts, that they have learnt the use of firearms, but now they possess a large number of muskets, most of which are old flint-locks, of English manufacture, bearing the Tower mark of various date, some as far back as the middle of the last century. The stocks of these are highly varnished and ornamented with red paint.[58]

As traffic in arms soon developed and the use of guns became widespread throughout the whole of Lushai Hills. The relative value of a gun can be seen from the fact that the Lushai raids into the plains during the second half of the nineteenth century was mainly to capture slaves, whom they sold to Pawis in exchange of firearms.[59] At this time, a slave of about four and half feet told was exchanged for two muskets.[60] In fact, Lushai regarded a brave patriot in high esteem, and in no way a person without guns could achieve such position. Hence, it became indispensably necessary to procure guns, as each and every male member of the society became anxious to possess muskets of their own.

---

56 *Ibid.*

57 *Ibid.*

58 R.G.Woodthorpe, *op.cit.*, p. 73.

59 F.Lalremsiama, *op.cit.*, p.4.

60 Alexander Mackenzie, *History of the Relation of the Government With the Hill Tribes of the North East Frontier of Bengal*, (Calcutta, 1884), p.296.

1
ANNEXATION

For about a decade following the Expedition of 1872, the Lushai frontiers remained relatively quiet.[61] A chain of outposts stretching from southern Cachar through Tipperah down along the Chittagong Hill Tracts and linking with those in the Arakan Hills had been established after extensive surveys and consultations between the military and the civil authorities on the defence of the frontiers.[62] Developments

---

[61] FPEP. February 1884; Nos. 302-309, K.W. No.1. Demi Official letter from A Mackenzie, to the Foreign Secretary. The peace was actually broken when on 18 January 1882, Hausata, in collaboration with the Muallianpui Chiefs, raided on the village of Lalsiva, the successor of Rothangpuia, Chief of Thangluah, for the latter's assistance to the British during the late Expedition. The raiders killed as many as twenty nine people, wounded seven and carried off ninety nine persons as captives and the raiders had taken number of animals like *gayals*, fowls, etc. The Muallianpuis attacked a Frontier Police party who were on the way to Demagri from Burkhal by Kolodyne river on the 18 November 1883. This daring raid was actuated by the famine conditions in the hills and was therefore clearly intended to plunder.

[62] H R Browne, *The Lushais, 1878-1889*, (Shillong, 1889, Reprint, Calcutta, 1978), P.4. See also Colonal E B Elly, *Military Report on the Chin-Lushai Country*, (Simla :1893, Reprint Calcutta ,1978), p. 11. "After the Expedition of 1871-72 the question of Frontier defence was reconsidered, and in 1874 Posts were established on the Cachar and Sylhet, frontier at Jirighat, Mainadhar, Monierkhal, Noarbund, Jhalnacherra, Chattachura, Oliviacherra, Lakhimamla, Fatehkuli, and Adampur. In Tipperah at Dharmanagar, Koilashut (Chirakuti), and Kamalpur. In Chittagong district at Demagri, Sirthay (Sirte), Ohipum(1) (Uiphum), Ohipum (2), Saichul (Saichal), Sungoo Valley, and Politai. In Arakan at Kulukwa and Daletmai. The Frontier Officers were also endeavour to keep up frontier relations by means of bazaar and annual meetings with Chiefs, at which feast were held and presents given before 1871 at Kassalong, and afterwards at Demagri."

within the Lushai Hills equally proved to be an effective check upon disturbances in its frontiers. Intensive rivalries between the eastern and western chiefs, which had broken out in open hostilities, had only served to absorb their warlike energies. As one officer with a later Expedition was to write:

> The lesson taught by the Expedition of 1871-72 appeared to have made a considerable impression upon the Lushais, for although, in pursuance of hereditary feuds, or in retaliation for recent insults and injuries, they still continued to raid upon each other, such disturbances were, for a long time, confined to their own territory.[63]

Even as these feuds showed signs of abatement pressure from the advancing Pawis and Lakhers diverted attention southwards.[64] The assistance given to the Lushais during the famine in the early eighties by the Cachar authorities, too, had brought in a measure of good will between them. But events at the close of the decade, both inside and outside the Lushai hills, however, were to bring these extensive hills under colonial rule but not before overcoming a protracted resistance.

FRONTIER DISTURBANCES

The Expedition of 1871-72 had not touched the southern hills and the 'Shendus,' as the Pawis-Lakhers were then known, were left very much to themselves. Inter village

---

[63] Surgeon Lieutenant Colonel A.S.Reid, *Chin–Lushai Land*, (First Edition 1893, Reprint Calcutta, 1976), Ch. IV, P. 58.

[64] Of the impact of such pressures on the British frontiers an official document thus stated: "a general movement would seem to take place from time to time amongst these people, apparently as if swarms were thrown off from the more crowded villages in the higher central hills such swarms forming new communities all around the outer fringe of the tract, and in doing so driving before them the villages which had previously inhabited this fringe. The inhabitants of them are compelled, in consequence of the pressure, to take refuge in our territory or in Tipperra or Manipur. There they are often followed themselves killed or taken captive and their villages plundered by the new comers. This seem to have been the origin of what is called the Kooki rising of 1849 and 1850, as it certainly was of the great series of raids in 1860-61." Quoted in B C Chakravarti, *British Relations with the Hill Tribes of Assam since 1858*, (Calcutta, 1964) p.49.

warfare, not merely one village against another, but often a combination of villages against another grouping went on relentlessly. Large parties on the warpath were often heard prowling about the Chittagong frontier.[65] It was the chance encounter between one such party and a British survey team that really marked the beginning of the frontier troubles.

After the annexation of Upper Burma of which the Chin Hills formed a part, the Government of India saw the need for a direct line of communication between and the newly acquired territories. As a preliminary to opening up communications between the two regions, Lieutenant John Stewart of the 1st Battalion, Leinster Regiment, was to begin a reconnaissance to the south east of Rangamati in the Chittagong Hill Tracts, to make a road and strengthen the frontier by additional outposts.[66]

He was to use Rangamati as his base and from there work southwards to effect a junction with another survey party under Lieutenant J Mc D Baird of the 2nd Battalion, Derbyshire Regiment.[67] Two men of his own Regiment, Lance Corporal Mc Cormick and Private Owens, to assist him in his survey operations, accompanied Stewart. His escort consisted of one *naik* and ten Gurkhas from the Chittagong Frontier Police together with some porters.

Stewart left Rangamati on the 16 January 1888.[68] But no sooner than he set out he received information from Demagri, supplied by the friendly chief Seipuia, that a large body of men from "Muallianpui, Thlantlang and Lungtian were on the warpath and proceeding in a westerly direction."[69] While eighteen miles off Rangamati on 3 February, Stewart received

---

65 C.E.Buckland, *op.cit.* p.180.

66 A.S.Reid, *op.cit.*, p.39.

67 *Ibid.*

68 FEAP. March 1888; Nos. 332-339. David Robert Lyall, Commissioner of the Chittagong Division to the Chief Secretary to the Government of Bengal. 11 February 1888.

69 *Ibid.*

another warning about raiding parties. But, as Colonel Scott Reid, the officer quoted earlier, wrote:

> With such contempt did he regard either the probability of an attack or the valour of his enemies, that when pressed by the *Naick* of the escort to be on his guard and to take additional precautions he replied: '*Kuki log ane se hamara salam do*' (give my compliments to the hill men and let them come on), and persistently declined even to post a sentry. He had left five *sepoys* of his small guard behind him, one to run with his elephants, another to take back his letters and three to look after part of the provision and form a depot a few miles to the rear.[70]

Lieutenant Stewart paid for all this with his life. Early next morning he was attacked and killed.[71] Stewart was apparently the victim of a quarrel between Hausata and his wife,[72] but it was an act that no government could overlook

---

[70] A.S.Reid, *op.cit.*, p.40.

[71] *Ibid.* p. 41. Reid thus described what happened: "In the early morning Lieutenant Stewart awoke, dressed, and gave the usual orders for the move, but had not left his hut or put on his boots. The two soldiers were still asleep in their own quarters, and the *sepoys* of the escort were beginning to move about, one lighting a fire, and the other engaged in the various preparations which precede the striking of a camp in the east." When the attack came, "Lieutenant Stewart received a bullet in the chest and fell back dead, not, however, before he and the two men had accounted for several of the enemy. Such was the story told by the survivors, but, as Stewart's body was afterwards found at the foot of a precipice twenty or thirty yards from the huts, some of the particulars are probably inaccurate."

[72] FEAP. October 1888; Nos. 87-117. C.S.Murray, Superintendent of Police, Chittagong Hill Tract to Inspector General of Police, Bengal, Panji Ham Dubashi's report of the enquiry on the outrage thus concluded: "Hausata's wife is a daughter of Jahoota; Hausata and his wife had a quarrel, and he beat her and generally ill used her. She then left her husband's home and went to her father, Jahoota, to claim his protection. Hausata followed her to Jahoota's village and tried to recover her. Jahoota at first refused, but finally agreed to give her up if Hausata would bring him two heads of any foreigners, i.e., of any one not a Kuki. Hausata agreed to this and assembled his warriors to the number of 150 and started off taking the following route: - They crossed the Phairuang kai, the Lungrang Hill, the other side of Uiphum range and made for the big Saichal, Hausata accompanying them. Seipuia got this information from his brother Vandula, who said to him, 'You have sworn an oath of friendship with the foreign chief Murray *Liena*. You should therefore warn him about this."

and called for the customary punitive expedition. Besides, it was evident that the existing system of the frontier outposts was powerless to check raids and that only the appearance of a British force in the Lushai country itself would prevent the tribes from creating trouble in the frontier.

Owing, however, to the lateness of the season, it was considered advisable to delay an expedition until the following cold weather. It was on 19 December 1888, that the Government of India issued orders for an expedition to the Lushai Hills. Its object were: to avenge the death of Lieutenant Stewart, to construct a good road to link Chittagong with Burma, and establish outposts which would dominate the tribes and prevent any hostility to their advance.

Even as the Government of India was working out plans for the expedition another serious incident took place. On 15 February 1888, an atrocious raid occurred upon the Chakma Roaja Prenkhyn Mro in the Chema valley in the Chittagong Hill Tracts, wherein six persons were killed, two wounded and twenty-three carried off as captives.[73] The raiders looted the village but strangely did not carry off the stored grain. Nor were the heads of the slain carried, indicating that it was in all probability a slave-raiding affair.[74] Until the expedition was set on foot the frontier outposts were strengthened by the addition of 250 men from a Bengal Infantry regiment.

Meanwhile, to the west of the Tuirial river significant developments were taking place. The death of the old British ally Suakpuilala in early 1881 was the signal for renewal of feuds between the western and the eastern chiefs. The first incident occurred shortly after Suakpuilala's death when a party from Pawibawia, Lalburha, and Chinhleia surrounded

---

[73] H K Barpujari, *Problem of the Hill Tribes: North East Frontier. (1873-1962), Vol. III*, (Calcutta, 1981). P.88.

[74] FEAP. March 1888; Nos. 322-339. D.R.Lyall to the Chief Secretary to the Government of Bengal. 25 February 1888.

the village of Thanghulha, but on the latter's submission no bloodshed occurred. A similar feud took place when Liankhama and his party killed the men of Thatliana and carried off forty heads. However, hostilities ceased when famine in the hills broke out.[75] Once these difficulties were overcome disturbances returned to these parts. The strong hand of Suakpuilala was no longer there to restrain inter-village rivalries.

On 13 December 1888, the Chakma chieftainess Pakuma Rani's village, only four miles off Demagri became the victim of another raid. The chieftainess and twenty-one men were killed, thirteen heads were taken, and fifteen persons were carried off as captives.[76] A small party from Demagiri followed the raiders but were unable to overtake them. A second party also failed to meet the raiders who had retreated too far to be caught.

The initial reports said that the raiders came from Suakpuilala's son Kalkhama's village, headed by Kalkhama himself. Survivors of the attack were said to have heard his name being shouted as the village was plundered.[77] The local authorities who took the raid rather seriously, was at a loss to explain why it had occurred. One view was that since there appeared no grounds for a quarrel between Kalkhama and the Rani,[78] its object was to obtain loot. The raiders could not have been mere headhunters.[79] The fact that the seven severed heads were badly mutilated, but not carried away also pointed to this fact. Another opinion suggested that the raid occurred on account of the chieftainess' pro-British antecedents. Pakuma Rani was a near relative of Rothangpuia,

---

[75] *Ibid*. pp., 86-87.

[76] FEAP. March 1889; Nos. 1-85, John Ware Edgar, Chief Secretary, Government of Bengal to the Secretary to the Government of India, Foreign Department. 18 December 1888.

[77] *Ibid*.

[78] *Ibid*.

[79] *Ibid*.

the British ally, who was much disliked by the Lushais due to his closeness to the British and particularly for his assistance to them during the expedition of 1871-72. Pakuma Rani too was close to the British and had always been friendly towards them.[80] Rothangpuia was long dead but his son and successor Lalchheuva's village was earlier raided upon. Likewise Pakuma Rani was supposed to have incurred the wrath of the other chiefs.

To find out the identity of the raiders the Chittagong authorities sent a number of men to the hills to make an enquiry. Till two months after the raids nothing certain was known of those involved or the village from which they came. Much less was known of their intentions. On 8 February 1889, however, David Robert Lyall the Commissioner of Chittagong Division and his assistant Murray met one of the captured women who with her child managed to escape and return to her village. Her name was Suaki, wife of one Vana, who said she was taken by Thata, *mantri* of chief Lungliana, in whose house she remained. From her statement it appeared that the actual leaders of the raid were Nikhama and Vanhnuna. Two other chiefs, Lungliana and Kairuma, the latter also from the same area, were also with them. She further added that there were sixteen prisoners of whom nine were in Lungliana's village and seven in Nikhama's. One girl was sold by Lungliana for sixty rupees to some Pawis in the south.[81] The women's account was confirmed by Hari Charan Sharma, political assistant to the Deputy Commissioner of Cachar. Sharma reported, on the authority of an old Lushai of Thanruma's *Punji*, named 'Singa Pa'* that Pakuma Rani's village was raided by Vuta's sons Lungliana and Kairuma. He had met

---

[80] *Ibid.*

[81] FEAP. March 1889; Nos.407-431. Lyall, Civil Political Officer. Lushai Expeditionary Force to the Chief Secretary to the Government of Bengal.

*More correctly Chhinga *Pa*, i.e, father (*pa*) of Chhinga, the form used in Lushai to refer to a person using the name of the first male child – "father of so and so."

Kalkhama who pleaded that neither he nor his brothers were involved.[82]

Even as an enquiry into the plunder of Pakuma Rani's village was proceeding news was received of extensive raids on Chakma villages in what was called the Chengri valley in the Hill Tracts during 8-10 January 1889. The first village to be attacked was Poojgong, which occurred in the morning of the 8th. According to the statement made by the chief of this village the raiders were about four or five hundred in number, who knotted their hair at the back of their heads, clearly indicating that they were Lushais and not Pawis or Lakhers.* The chief's wife, his two daughters and a daughter-in-law were carried away after most of the houses in the village were burnt or razed to the ground.

The next day, at about four in the morning, Lotibong village was attacked. This was followed by the plunder of other villages. During the three horrifying days, no less than twenty three villages had been plundered.[83] The method used in these raids was somewhat novel: the raiders separated themselves into small groups of twenty to thirty fighting men attacking different villages simultaneously. Once this was over they regrouped and proceeded home with all their captives. The total casualty in the Chengri valley affair was 101 persons killed and two wounded. The numbers of captives were about 150, among whom were 100 women and the rest children below fifteen years. The total number of houses destroyed were 545 and the value of the property looted or destroyed was estimated at Rs.18,582/-.[84]

The Chengri valley raid was attributed to Zahrawka, younger son of Suakpuilala and younger brother of

---

82 *Ibid.*, No. 426. J.Kennedy, Deputy Commissioner of Assam to the personal Assistant to the Chief Commissioner of Assam.

*The Lushais of the Lusei clan used to knot their hair at the back of their heads, whereas the Pawis and Lakhers tied their knots on the top.

83 L W Shakespear, *History of the Assam Rifles*, (London, 1927), p. 87.

84 FEAP. March 1889; Nos. 407-431, see Statement of Singbir Thapa.

Lianphunga. At this time Chengri valley was not under the control of either the British or the Tipperah Raja. Zahrawka laid claims to it as his elephant hunting ground. Apart from this the inhabitants were said to be often hanging about Zahrawka's *jhum* cultivation and scaring the women and children who worked in the fields. Advised by his elder brother Lianphunga, Zahrawka raided the Chengri valley villages. Lianphunga himself accompanied him in this raid.[85] Zahrawka and Lianphunga refused to release the captives to messengers sent by the district authorities in Cachar. Instead he insisted upon ransom money which was said to be due to him from some earlier transaction.[86]

LUSHAI POLICY : RADICAL CHANGE

Meanwhile significant developments were taking place east of the Lushai Hills. The incorporation of Upper Burma

---

[85] FEAP. October 1889; Nos. 27-66, J D Anderson Deputy Commissioner of Cachar to the Secretary to the Chief Commissioner of Assam. 13 July 1889. Lengpunga's own account of the Chengri valley raid appeared to be as follows : While Zarock ruled a separate village, he laid claim to the land of the Chengri valley people, and informed them that, in as much as they were not tributary either to the British Government, or to the Maharaja of Tipperah, he intended to assert his right to their land as an elephant hunting ground. Shortly after this, some of the Chengri valley Kukis hovered about Zarock's *jhum* cultivation and scared his women and children. Zarock took council with his brother Lengphunga, who (he candidly admit) advised him that it was his bounded duty to subjugate the Kukis, and for that purpose lend him some of his young men.

[86] *Ibid.* "... seven of the captives have been ransomed from villages situated in the vicinity of the Dallesari belonging to Zarock and Lengpunga, sons of Sukpilal. It is reported that Rs.700/- was demanded as the ransom for these captives, and that as the persons who went in search had only Rs.185/-, they had promised to pay the balance Rs. 515/-, within a month. It is said that Lengpunga has threatened, if this money is not paid in due time, to commit further raids and to murder the remaining captives supposed to be sixty six in number. It is not known where all these captives are, but there is reason to believe that other villages besides those of Zarock and Lengpunga were concerned in the outrages... It will be seen from the above that although we have not as yet been able to obtain full information regarding the perpetrators of these outrages, we know with certainty who some of them are, and approximately the position of their villages, which lie North and East of the line taken by us in advancing on the Shendoos and west of the tribes of Chins, who have been so successfully dealt with this year by General Faunce."

after the third and final Anglo-Burmese War (1885-86) brought about a radical change in the policy of the Government of India towards the Lushai Hills. Hitherto, the area occupied by the Lushai tribes formed the 'real frontier' extending up to the border of Burma when the occupation of these hills would have brought the British into immediate contact with the tribes then under 'the imperfect control.'[87] Thus the expeditions in the early years were no more than what was called on the North West Frontier as "butcher and bolt". No attempt was made at controlling the tribes from within.

The new policy of establishing control over all the hills in the vast region between Bengal and Burma had already begun in the Chin hills. The first step was the occupation of the Kale state south of the Kabo Valley, east of Manipur. At a *darbar* held on 26 March 1887 at Phongyikyang in the Kale state, the local chief Maung Pa Gyi was recognised as governor of that state, and the Chins were told that no raids upon the territory occupied by the British and no slave taking or trading would be permitted.[88] Slaves taken before British occupation were, however, permitted to be retained. It was also announced at that *darbar* that the British intended to open a trade route from Kale to Chittagong.[89]

For the occupation of the tract a chain of military posts was established in the area extending from Tamu and other stations in the Kubo valley to Gangaw, Minywa, and Tilin (Tiddim) on the Yaw border in the Chin hills. The country was surveyed, local resistance was crushed, and dacoity stamped out.[90] Next the British tried to give a crushing blow to the Siyins and the Tashons and release their Shan captives.[91] But Chin resistance proved more determined.

---

[87] H K Barpujari, *op.cit.*, p. 90.

[88] Vumson, *Zo History*, (Aizawl , 1986), pp. 113-114.

[89] *Ibid.*

[90] Bertram S Carey and H N Tuck, *The Chin Hills, Vol. I*, (Falam, 1932, reprint, 1983). p.20.

[91] Vumson, *op.cit.*, p.117

Khawcin, chief of Kamhau, group of Chins, rounded up 1400 fighting men, and held the British at a place called Khuasak. Another attempt was made by the Chin to stop the advance of the British force at Hanthanglu. Here the British established a strongly fortified position, known as Fort White. The establishment of similar strong posts within the hills and the movement of small flying columns crushing each village finally put an end to Chin resistance.[92]

The series of events in the Lushai hills border since the killing of Lieutenant Stewart and the progress of military measures in the Chin hills made it clear that any plan for dealing with the hill men should be worked in concert by the Governments of Bengal, Assam and Burma. This would entail making in the first instance a road through the Chin hills and Lushai hills to connect Chittagong with Burma. John Ware Edgar, the Chief Secretary to the Government of Bengal thus wrote to the Government of India, about his Government's views:

> ...no mistake can be greater than for each of these governments to deal separately with the villagers adjoining its own frontier without reference to those in the vicinity of the other two governments, or in the centre of the tract, because the effect of this would be merely to divert attacks from one portion of the frontier to another while doing nothing to remedy the real source of the evil which I take to be the belief of the inhabitants of higher central hills in the inaccessibility of their country and their safety from the danger of punishment.[93]

Edgar presumed that the frontier disturbances might be put to an end to if the governments of the surrounding country unitedly applied the same methods of dealing with these tribes. The Lushai hills were now surrounded on the one side by the

---

92 For details see Carey and Tuck. *op.cit.*, see also Sir Charles Crosthwaite, *The Pacification of Burma*, and Dorothy Woodman, *The Making of Burma*,

93 Edgar, to the Secretary to the Government of India, Foreign Department, 17 December 1888, quoted in H.K. Barpujari, *op cit.*, P.89.

settled districts of Bengal and on the other by the Chin hills over which control was being established. The Governor General in Council accepted Bengal's proposal for an expedition and it move into the hills. The purpose of the operations, it was made clear, would be mainly to open the communications between Demagri to Seipuia's village further east and then to move further on to punish Hausata's village.[94] This punitive expedition was to be undertaken with reference to Lushai raids. The manner in which the country should be dominated was yet to be decided by the Governor General in Council and Bengal was accordingly told that it would be communicated to them hereafter.

OCCUPATION OF LUNGLEI

The Lushai expeditionary force under the command of Colonel Vincent Tregear consisting of 250 men of 2nd Bengal Infantry, 250 men of 9th Bengal Infantry, 400 men of 2nd Battalion 2nd Gurkha Regiment, 200 men of 4th Madras Infantry (Pioneers), two guns of No 2 Bombay Mountain Battery with 1000 coolies, moved to Demagiri and then to Seipuia's village on 17 February 1889.[95] By mid March a road up to Lunglei was completed by Captain John Shakespear, where a large timber stockade was built and came to be known as "Fort Lunglei". John Edgar, in his letter to the Secretary to the Government of India had thus reported the progress of the expedition:

> A good hill road, practicable for elephants and mules has been made from Demagiri to Lungleh, a hill 3500 feet high in the neighbourhood of the village at present by Saipuya. At Lungleh a strong post had been fixed, which will be held during the rain by a force of 200 of the Frontier Police.[96]

A reconnaissance party along with two guides provided by a chieftainess, Darbilhi, a near relative of Hausata, set out

[94] *Ibid.*, No.28.

[95] A.S.Reid, *op.cit.*, p.52.

[96] FEAP. October 1889; Nos. 27-66.

on the operations against the latter on 14 March 1889. The force visited Zahuata's village first on the 19 March and proceeded towards Hausata's though information had already been received that the chief was dead. The action in Hausata's village thus described by one of the participants:

> At 9-10 A.M. we got within 800 yards of Hausata's village where we could see a few men. The ridge here turns at a right angle, and we could see across the ravine. The men called out something, but neither the guides nor Mr. Murray could understand what they said, as they were a long way off. A shot fired, clearly as a signal to burn the village, for almost immediately flames broke out in two or three houses in the lower end of the village. Colonel Tregear at once ordered up the gun, and also directed some volleys to be fired. The practice was good, and at once stopped the further firing of the village.[97]

The inhabitants fled after sacrificing a dog and leaving its remains across the path approaching the village. When the troops entered the village they found it totally deserted, and even a stockade that had been constructed for their defence was abandoned.[98] The guides from Seipuia village were understandably reluctant to show the expeditionary force the grave of Hausata. On being persuaded to do so, the grave was opened and beneath the chief's dead body, Stewart's gun was found, clearly proving that the dead chief was unmistakably the man behind the attack upon that officer and his party.[99]

From Fort Lunglei as his base General Tregear scoured the country side. Thereafter on 3 April 1889, he held a *darbar* of the southern chiefs at Lunglei. The three great Haulawng chiefs Seipuia, Lalthangvunga and Lalluauva, all of them Vandula's brothers along with Vandula's sons Sangliana and Lalthuama were in attendance. Vandula was by then an old

---

[97] *Ibid.*

[98] *Ibid.*

[99] FEAP. October 1889; Nos.27-66, Lyall, Commissioner of Chittagong division, to the Chief Secretary to the Government of Bengal. 5 May 1889.

man, and did not attend, but not so much because of his advanced years as for his dislike of the British. General Tregear told the assembled chiefs to help him in his further operations against the offending chiefs to which the chiefs gave their assurance.[100]

Owing to the lateness of the season the Pawi and Lakhers, or Shendus were not effectively dealt with the exception of the burning of the village of Hausata and Zahuata. None from these villages attended the *darbar.* At fort Lungleh a garrison of one British officer and 212 men of the frontier police was locate and the road linking Demagri to that place was completed.[101] By 16 April the main force returned to the base leaving Murray to command the Fort.[102]

CHIN-LUSHAI EXPEDITION

In reviewing the results of the expedition Bengal's Lieutenant Governor felt that to effectively deal with the Shendus and to punish the raiders of Pakuma Rani's village and Chengri valley and to release the captives taken, a bigger expedition into the hills was called for. The operations in the Chin hills had to be coordinated with those proposed from the Lushai hills.[103] The Government of India once again accepted Bengal's proposal and sanctioned the combined expedition on 11 September 1889, which came to be styled as the "Chin-Lushai Expedition."[104] Its aim was to subjugate the tribes as yet neutral, to explore and open out, as far as practicable, the partially known country between Chittagong and Burma, and to establish semi permanent posts in the tracts so as to ensure complete pacification and recognition of British power by the tribes.[105] The whole expedition was placed under General William Penn Symons General Officer

---

[100] L W Shakespear, *op.cit.*, pp. 88-89.

[101] Sir Robert Reid, *History of the Frontier Areas Bordering on Assam*, (Shillong, 1942, reprint, 1978), p. 11.

[102] FEAP. July 1895; Nos. 122-145.

[103] Reid, *op.cit.*, p.11.

[104] *Ibid.* p.14.

[105] H K Barpujari, *Problem*, *op.cit.*, p. 92.

Commanding Migyutin, with immediate control over the Burma forces. Vincent Tregear, now a Brigadier General, was to command the forces operating from the Chittagong side.

The Chittagong Column, consisting of 3977 men including Captain Shakespear as Field Intelligence Officer and Colonel G.J.Skinner, with C.S. Murray as Political Officer,[106] that moved from the southern Lushai hills started from Demagri under the command of General Tregear in November 1889. The work entrusted to it included the construction of a mule path to Haka in the central Chin hills to open up communications from Bengal to Burma; to establish a post in some suitable place; and to subjugate the tribes and bring them under British control. From this main force a column was to march towards the north to punish Lianphunga for his complicity in the Chengri valley raid and Vuttaia's sons for their raid in Pakuma Rani's village, and thereafter to establish a permanent post near Lianphunga's village.[107]

Chittagong Column reached to Lunglei in a matter of weeks. From there reconnaissance parties were sent in various directions. Brigadier General Tregear arrived at Lunglei on 24 December and thereafter a further advance eastward was made, across the Bualpui range and down to the Mat River and finally to Chhimtuipui. From there they ascended to the Darzo hill, where they found a good open space. As this place was good for the heliographic communication with various reconnoitring parties a strong defensive post of 200 rifles was established and came to be known as "Fort Tregear".[108]

On 13 January fifty men of the 2nd Gurkhas under Captain Hall and the same number of Frontier Police under Murray accompanied by Captain Shakespear proceeded from Lunglei to explore the country between the Mat and the Tuichang rivers and to find out the real disposition of the chiefs of that

---

106 Reid, *op.cit.* p. 10.
107 A.S.Reid, *op.cit.*, pp.191-192.
108 L.W.Shakespear, *op.cit.*, p. 92.

region, of whom was included Lathuama.[109] A Pawi chief Paawna, who was reported to have been involved in the murder of Lieutenant Stewart, was found sheltered by Lalthuama. On reaching Lalthuama's village two days later, a fine of 10 pigs, 10 goats, 20 fowls, 30 guns and 100 baskets of grain on the chief was imposed.[110] Lalthuama was taken to Lunglei until the fines were paid up. Seven days later this was done and he was accordingly released.[111]

A flying column under Captain Shakespear and Mulaly with fifty rifles started on 2 February from the Mat river camp to visit the Muallianpui chieftainess Darbilhi, who had maintained a cordial relationship with the British since the Lushai Expedition of 1889. The objects of the expedition were to reassure friendship, obtain the submission of the Muallianpuis, survey the country, and, if possible, open communications with General Penn Symons' column by heliogram.[112] The column reached Darbilhi's village on 4th where they were well received.

Messengers were sent to Vantura and Dokulha, both of whom were also reportedly involved in Stewart's murder, to inform them to come in. Vantura submitted himself and gave up Stewart's revolver which he said he had obtained from Zahuata. Pending the full payment of a fine of a thousand maunds of rice and hundred guns imposed upon him Vantura was kept in custody. In the meantime Dokulha pleaded illness and failed to come in. In addition to Darbilhi two other Muallianpui chiefs, Dokapa and Patlaia, took the oath of friendship and professed their willingness to assist the British. Thus the object of the visit to Muallianpui was almost completely attained. However, heliographic communication with General Symond's column was found impracticable on account of an intervening high range.[113]

---

[109] Lalthuama was the youngest son of Vandula, who was the head of all the southern Haulawng and the son in law of Lianphunga.

[110] A S Reid, *op.cit.*, p.194.

[111] *Ibid.*

[112] *Ibid.*, p. 197.

[113] *Ibid.*

On 24 February a reconnoitring party under Captain Hall left Darbilhi camp towards Haka. Two days later they met Captain Rundall and his party who had marched from Haka. The meeting of the two forces represented the "laying of the keystone in the arch," which now stretched across the wild country separating Burma from India.[114] The combined parties proceeded to Haka from the Tiau river on 28 February. The following morning they reached Tlualam village where they met chief Lalhleia who confessed that Stewart's aneroid and field glasses were in his possession and that one rifle was in Tlualam and the heads of the victims and other rifles were in Paawna's village Tangzang. Meanwhile some heads and one rifle were brought in from Zahuata's village to Captain Shakespear. The following day Shakespear marched towards Thlan Tlang where the aneroid and field glasses were recovered and the surrender of a Burmese captive was also received.[115] At the next camp, in the valley of the Sa the barrel and the lock of a Martini-Henry rifle was brought from Tangzang village to Captain Shakespear which were said to have been brought by orders of Lalhleia. The next march took the parties into Haka on 3 March.

After three days halt Captain Hall returned towards the Lushai hills from Haka. On the 7th he reached Tlualam where he found a captive Lushai woman. But their chief Lalhleia was said to have submitted to General Penn Symons and the *mantri* refused to surrender her without his authority. When the party reached Dopawrha's village a female captive taken from Lalchheuva's village and one who had been sold to chief Patlaia was surrendered. On the 26 February seven more captives from Hausata's village were also surrendered by Vantura.

By April–May 1890 the expedition succeeded in releasing captives, imposing fines and generally punished chiefs guilty of resisting the British or creating disturbances. Roads had

---

114 *Ibid.*
115 *Ibid.*

been built and telegraph lines laid out. Colonel A S Reid thus summed up the work of the Chittagong column of the Chin-Lushai Expedition:

> A good road of easy gradients had been made from Fort Lungleh to Fort Tregear and beyond the advanced post a fair mule path led to Haka 81 miles, completing the connection of India and Burma. Two large rivers Mat and Kolodyne, had been bridged. These bridges, the former 206 and the latter 304, feet long had to be very substantially built, and were of the crate and trestle type.[116]

OCCUPATION OF AIZAWL

In order to punish Lianphunga for his raids upon the upper Chengri valley, the Northern Lushai Column as it was called, under the command of Colonel G J Skinner, accompanied by Murray as Political Officer, was detached from the main force of the Chittagong Column. The task of punishing the sons of Vuttaia for their raid on Pakuma Rani was also assigned to this column.[117] With the help of a guide provided by the Haulawng group, the force left Lunglei by the end of January 1890 and moving up[118] the Tlawng River as far as it was possible for rafts, reached a point near Lianphunga's village.

---

[116] *Ibid*,. pp. 207-208.

[117] FEBP. March 1890; Nos. 45-56, Diary of events, Chittagong Column, Chin-Lushai Expedition, for the week ending the 2 February 1890. The Northern Column consisted of Colonel G.J.Skinner, 3rd Bengal Infantry, Commanding, Lieutenant H.W.G.Cole, 2nd/2nd Goorkhas, Staff Officer, Captain O.A.Chambers, Warwickshire Regiment, Intelligence Officer, Lieutenant W.H. Birkbeck, 1st King's Dragoon Guards, Signalling Officer, Surgeon J.C.Lamot, in charge of A section, No.9, Field Hospital, Captain W.H.Allen, D.A.C.G, Transport Officers:-Captain E.Montague, Sufflok Regiment, Lieutenant T.St. C. Davidson, Leinster Regiment, Lieutenant E.H.S. Twyford, Scottish Rifles, Lieutenant M.C. Willoughby, 2nd Bengal Cavalry, with C.S.Murray, Assistant Political Officer, *Sappers and Miners* :- Captain R.D Petrie, Lieutenant E .H. de V.Atkinson, 1 British Non-Commissioned Officer, 80 Native Ranks, *3rd Bengal Infantry*:- Captain F.B.W. Richardson, Lieutenant Warren and Wikeley, 250 Native Ranks, *2nd/2nd Goorkhas* :- Major F.R.Begbie, Captain F.P. Hutchinson, Lieutenant E.J.W.Drummond, Surgeon H.Fooks, 300 Native Ranks, *28th Bombay Pioneers* :- Lieutenant G.A. Brownrigg, 102 Native Ranks, *Frontier Police* :- Fifty Native Ranks.

[118] L.W.Shakespear, *op.cit.*, p.90.

Meanwhile, a simultaneous advance was made from Cachar by 400 men of the Surma Valley Military Police,[119] under the command of W W Daly with the order to recover the captives in the Chengri valley raids and to punish Lianphunga, and establish a stockade and godown at Changsil, collect information, and effect a junction with Skinner's Column under whose command he was to operate.[120] The movement of this force was ventured in close consultation with the Governments of Assam and Bengal.[121] Daly reached Aizawl on the 4 February 1890, where he built a stockade on a commanding site. "Fort Aizawl" as it came to be called, and eventually became the headquarters first of the North Lushai Hills and later the united Lushai Hills District. From here he advanced to Lianphunga's village on 8 February and persuaded Lianphunga to come to terms, on the promise that he would not be detained and his life would be spared.[122]

With the arrival of Colonel Skinner on 11 February, Lianphunga became nervous and fled into the jungle. Skinner and Daly occupied his village, only to be heavily attacked by the Lushais who also set fire to the houses in which their men were sheltered. A sortie or two drove off the assailants after which the village was destroyed along with over a thousand maunds of unhusked rice, its granaries and livestock. The Lushais were pursued into the jungle and hunted down mercilessly. In view of his assurance to Lianphunga before the arrival of Skinner's column Daly did not consider himself justified in attempting to seize the chief, at least not until he received further orders from higher authorities. Lianphunga had already surrendered sixty three captives, all women and children, and Daly had taken care of them well enough.[123]

---

[119] *Ibid.*, It included 10 British Officers, J.R.Carnac, L.St.Brodrick of the Assam Police, S.N.Walker of the Bengal Police, with Dr.Partridge and Surgeon Coleman of the 43rd Gurkha. Rifles, as Medical Officers.

[120] Reid, *op.cit.*, p.10.

[121] *Ibid.*

[122] A G McCall, *Lushai Chrysalis*, (Shillong, 1942, reprint, 1977), p.58.

[123] A.S.Reid, *op.cit.*, p.217.

Nevertheless as a result of the destruction of Lianphunga's village all the chiefs up to the Fort Lunglei one by one began to tender their submission. Kalkhama and seven of his kinsmen, all sons of the late chief Suakpuilala, too tendered their submission.[124] But they refused to surrender Lianphunga, and it was clear that they were ready to face the consequences. It was only because of General Tregear restraining him that Colonel Skinner, angered by their refusal to surrender Lianphunga, was prevented from burning all their villages. [125]

To punish the sons of Vuttaia for their raid upon the village of Pakuma Rani, a flying column of nearly two hundred men under Major Begbie[126] advanced upon the villages of Nikhama and Lungliana on 6 March. Both villages were occupied and the chief took to the jungles. Lungliana's *mantri* surrendered six captives but a considerable number remained. On the failure of the chief to pay the fine and surrender all captives within the given time, his village was burnt. Major Begbie thus inflicted a fine of sixty guns and three hundred rupees on Nikhama to be realised within twenty four hours. He was also ordered to surrender all captives.[127] On his failure his village was also burnt on 9 March 1890.[128] Thereafter the force returned to Lianphunga's village on the 12th

By the close of March the Chin-Lushai Expedition had successfully accomplished its tasks. Those responsible for the disturbances in the Chittagong Hill Tracts were adequately punished. A mule track from Bengal to Burma was put into operation.[129]

---

124 FEBP. May 1890; Nos.1-7, Dairy of events, Chittagong Column, Chin Lushai Expedition, (Northern Column's report) for the week ending 10 March 1890. Reported by Brig. General V.W. Tregear.

125 *Ibid.*

126 A S Reid, *op.cit.*, p.220.

127 *Ibid.*

128 *Ibid.*

129 *Ibid.*

ADMINISTRATIVE ARRANGEMENTS

In the north was established Fort Aizawl with a garrison of 200 military police and its subsidiary stockade at Changsil with another 100 men.[130] Changsil and Aizawl were connected by a fair road but no other means of communication with Silchar existed beyond that afforded by the river Tlawng; the telegraph line from Changsil to Jhalnacherra was only partially completed.[131]

On the administrative side, the operations led to the creation of two districts–North Lushai Hills and South Lushai Hills. In connection with the occupation of the Lushai Hills, J W Quinton, Chief Commissioner of Assam on 15 May 1890 remarked:

> Mere occupation by a police force of certain points in the tracts referred to would not in itself be sufficient to bring under our influence the Chiefs with whom we have been so lately in collision, and that, if this object was to be adequately attained, it was essential that an Officer possessing both experience and judgement should be at the same time appointed to feel his way among the people and gradually accustom them to our control.[132]

Quinton further added the necessity for a separate administrative setup thus:

> It is true that our intercourse with the Lushais has hitherto been conducted under the control of that Officer, but such intercourse has been only very slight, and our dealings with the Lushais have been few and infrequent. This state of things has, however, been completely changed by the late expedition, and the Deputy Commissioner could not, consistently with the due performance of his other duties, spare the time required for the closer control and more constant communication, from which alone the extension of our influence over the tribes concerned can be looked for.[133]

---

[130] *Ibid.*

[131] *Ibid.*

[132] Reid, *op.cit.*, pp. 19-20.

[133] *Ibid.*

With the conclusion of the Chin-Lushai operations and the reduction of the Lushai Chiefs to submission, the question of the future administration of the newly acquired tracts began to engage the attention of the Government of India. Such details as whether there should be one head of the civil administration and one military commander, or whether these duties should be combined in a single officer, the civil staff to be employed, the number of districts into which the new tracts should be divided, their boundaries, the organisation and distribution of the military, arrangements for transport and commissariat, and the political question of the retention of the chiefs and the kind of taxes and tribute to be imposed on the people had to be decided. Pending a settlement of these questions, tentative administrative arrangement was made for the occupation and control of the Chin-Lushai Hills.

For administrative convenience, the Chin Hills already under Burma was divided into two tracts, viz. The Northern Chin Hills controlled from Fort White and the Southern Chin Hills from Fort Haka. Bertram S. Carey and Captain Ross were appointed as Political Officers respectively.[134] The Chin hills remained under the direct administration of the Chief Commissioner of Burma.

The southern Lushai Hills, from now on referred to as South Lushai Hills, was attached to Bengal under its Lieutenant Governor from 1 April 1891. C.S. Murray of the Indian Police was appointed as the Superintendent with the Headquarters at Lunglei.[135] The district was put under the control of the Bengal Frontier Police, pushed up from the Chittagong Hill Tracts, which consisted of 529 native Officers, non-commissioned Officers and constables.[136] Besides Lunglei, Fort Lalthuama and Fort Tregear were established with a base at Demagiri. The boundaries of the district was in the west the Chittagong Hills District on which it abutted for its entire length, on the east lay the Chin hills while in the north

---

[134] FEAP. March 1890; Nos. 70-71.

[135] FEAP. September 1891; Nos. 38-59.

[136] *Ibid.*

and south its boundaries marched with the North Lushai Hills and the Arakan Hill Tracts respectively.

The northern region, the Northern Lushai Hills, was put under Assam from 3 June 1890. Captain Herbert Browne, the Personal Assistant to the Chief Commissioner of Assam, designated the Political Officer, was to exercise general control over all departments with headquarters at Aizawl.[137] The hills were to be held by the Surma Valley Military Police Battalion numbering 300 then at Aizawl which was soon increased by an additional 200 men. Posts were established apart from that at Changsil, at Sairang, Rengte road and Tuirial, all on the road to Cachar.[138]No definite boundaries were laid down. The district was limited to the areas held by the descendants of Lushai chief Lallula Sailo, that is the tracts lying in the Cachar frontier on the north, Hill Tippera on the west, Manipur in the east and an imaginary line east-west through the Darlung peak, and the South Lushai Hills in the south.[139] The Secretary of State for India accorded this sanction to the several arrangements and ordered that these might be allowed to continue till such a time when he was in a position to propose a change.[140]

---

[137] FEAP. August 1890; Nos. 221-227. Under the sanction conveyed in Foreign Department telegram No.1308 E of 25 June 1890, confirmed by the Secretary of State's Despatch No.64 (POL) of 18 September 1890.

[138] FEAP, October 1890; Nos. 355-357. The force was necessary not only for defensive measures but also to enable the Political Officer to promenade through the country which was needed to keep the Lushais under proper control.

[139] H.K.Barpujari, *op.cit.*, p.95

[140] FEAP. August 1890; Nos. 1-3, The Sub-division of the Chin Hills into two tracts was done after the suggestion of the Chief Commissioner of Burma who felt that the tract was 'too extensive' to be administered from one centre and as several tribes occupying the country remained unsubdued.

2

WESTERN LUSHAIS: REVENUE AND LABOUR

With the possible exception of Sir John Ware Edgar[141] who was Deputy Commissioner, Cachar, during the Lushai Expedition of 1871-1872, few men were better informed of the Lushais and the Lushai hills than Captain Herbert Browne. As Quinton's Personal Assistant in Shillong he had handled all the correspondence relating to the Lushai hills and had compiled a detailed hundred and eighty seven-page note on British relations with these people during the past decade.[142] He had inherited much of his personal knowledge and interest in the Lushais from his father, Lord Ullick Browne, who as Commissioner of Chittagong, some twenty years ago had greatly influenced Bengal's Lushai policy. He was therefore well qualified to carry out Quinton's ideas on

---

[141] C E Buckland, *Dictionary of Indian Biography*, (London, 1905, reprint, New Delhi, 1971), p. 131. Sir John Ware Edgar (1839-1902) ICS; born 16 September 1839; arrived in India in February 1862; service in Cachar, accompanied Lushai Expedition of 1871-72 as Political Officer of the Northern Column; as a Deputy Commissioner of Darjeeling he paid much attention to Sikkim, Buddhism and Tibetan frontier questions. President of Bengal Excise Commission, 1881-83; Chief Secretary to the Government of Bengal, 1887-1891; Member of the Governor General's Council, January 1892; retired in April 1892; CSI in 1872, KCIE in April 1892; devoted himself in later years to historical studies, chiefly on subject connected with Northern Buddhism and modern Latin Christianity; died at Florence on 4 June 1902.

[142] Herbert Browne, *op.cit.*, pp. 1-187.

the administration of the Assam portion of the Lushai hills. Before Browne left Shillong Quinton issued him general instructions on what he was to do in the North Lushai Hills:

> Your headquarters will be at Fort Aijal but you should keep moving among Chiefs with the object of establishing political influence and control over them, and inducing them to submit themselves gradually to our rule. As far as your means will allow, you should further endeavour to put down open raidings to protect our friends, and to punish those who injure them. You will otherwise not be strict to mark what is amiss or attempt to introduce a criminal administration, which, under existing circumstances, you are not in a position to enforce. You will leave the inhabitants, as far as possible, to settle their own affairs among themselves. For the present, you should consider as coming within the scope of your influence the tribes inhabiting the tract lying between the Cachar Frontier on the north, Hill Tipperah on the west, the Manipur river on the east, and on the south an imaginary line drawn east and west through the Darlung peak.[143]

Browne was to attempt opening communications with officers of the Bengal Government at Lunglei and Fort Tregear in South Lushai Hills and with those of Burma at Haka and Fort White in the Chin Hills. He was also to take every opportunity of procuring information regarding the numerical strength of the several tribes under his charge, particularly in such matters as tribal customs and organisation, which could be of use in the future. But the more important part of the Chief Commissioner's instruction related to Lianphunga's complicity in the Chengri valley raids. Browne was to conduct an enquiry and submit a report recommending appropriate punishment. On the equally important question of tribute or taxation Quinton enjoined caution:

---

[143] Sir Robert Reid, *History of the Frontier Areas Bordering on Assam*, (Shillong, 1942, Reprint , 1978). pp. 21-22. See also FEAP. August 1890; Nos. 238-262. F.C.Daukes, Officiating Secretary to the Chief Commissioner of Assam to Captain Browne, 22 April 1890.

> If, as the Chief Commissioner understands ... to be the case, some of the chiefs are willing or anxious to pay revenue or tribute you will receive it in money or kind as tendered, subject to the orders of the Chief Commissioner, and you will endeavour to induce others to follow their example; but no attempt should be made at present to exact revenue or tribute from tribes unwilling to pay it. You should report fully in due course what you find the position to be in regard to the willingness or otherwise of the tribes to pay revenue or tribute; and, pending the final orders of the Chief Commissioner, you should be careful not to accept, in return for payment of revenue or tribute, any obligations of a nature which might render their future fulfilment a matter of difficulty.[144]

The new Political Officer was also directed to reopen the market at Changsil, which the Lushais were said to be anxious about.[145] Once this was done the country between Changsil and British territory was to be examined so that a road could be opened connecting the Lushai Hills with Cachar.

DARBAR AT AIZAWL

Captain Browne reached Aizawl on 18 May and at once proceeded to carry out the policy chalked out for him. His first effort was to bring in all chiefs to a conference in Aizawl where a decision on Lianphunga and Zahrawka could be announced and future relations between the chiefs and the British indicated. Lianphunga himself was to attend. A *mantri* of Thanruma, a village situated about seven miles from Aizawl, was entrusted with the duty of informing all the chiefs that the *darbar* was to be held shortly. But the *mantri* informed Browne that Lianphunga had already left his village and that all the chiefs decided that Thanruma, in view of his proximity to Aizawl, would represent them.[146] Though unhappy that

---

144 *Ibid*. p. 22.

145 FEAP. October 1890; Nos. 238-262, Daukes to Captain Browne, 22 April 1890.

146 *Ibid*., (No.129) Diary of Captain Browne, week ending of 25 May 1890.

W W Daly had earlier refused to meet them the chiefs now agreed to come in about fifteen days.

Meanwhile enquiries had been made about the payment of revenue. The Political Jamadar Sib Charan had told Browne that the chiefs had never offered to pay revenue to Daly. Sib Charan and another employee were accordingly sent to Thanruma's village and on their return on 11 June reported that the Lushais told them they would never pay revenue. If payment of revenue was a necessary condition before their making friends with the British, there would be no friendship at all. Sib Charan also explained to Browne that few of the Lushais did anything but cultivate, a few collected wax and rubber, and got money that way, but the majority had no money or the means of getting it.[147] After Browne enquired of the Political Jamadar whether the Lushais will do coolie work, and earn money that way he was told that there would be no chance for it for another two or three years. Sib Charan added, "it was not their custom, and their chiefs being afraid of loosing power, won't let them."[148]

As directed by the Chief Commissioner, Captain Browne at the same time began to investigate the complicity of Lianphunga and Zahrawka in the Chengri valley raid. From the statement earlier made to Daly, on the 9 February 1890, it was clear Lianphunga had admitted his involvement in the affair.[149] He had told Daly that he did not know the unfortunate village was under the protection of the British. The area was their hunting ground, he had affirmed and:

> I was not aware that the village was in British territory, but I thought it was in Hill Tiperrah. Formerly, many

---

[147] *Ibid.*

[148] *Ibid.*

[149] *Ibid.*, (No. 133) Browne to Secretary to the Chief Commissioner, Assam. "Punishment of Lengpunga and Jarak for their complicity in the Chengri Valley raid." 14 June 1890.

> years ago, Savunga had a *punji* there, which was attacked by the Tiperrah and cut up. Afterwards that part of the country was used as hunting ground by the Lushais. When the Chengri valley people settled there, about the year of the great earthquake (1869), we told them not to remain, as it was our hunting ground, and that if they persisted, we should attack them.[150]

Lianphunga on the other hand, had his own complaints: he had helped in the release of the captives, but the officer who obtained the release of eight captives by borrowing money from him had not repaid the loan till then. He had also stated that he told two subordinate officials who came to claim the release of other captives that these were not British subjects; but that if they considered them as such they could take all of them. This was declined, Lianphunga had said, on the ground that the officials could not take so many men along with them and had instead entrusted them to his care.[151]

A *darbar* of chiefs was held at Aizawl on the 13 June. Captain Browne had decided to announce to the assembly the punishment of Lianphunga and Zahrawka, as the former had meanwhile repeated to him what he confessed to Daly. Browne felt that no further proof of the complicity of these chiefs was needed and accordingly told the assembly that he had decided to depose the two for a term of four years and to fine them fifteen guns; ten guns from Lianphunga and five guns from Zahrawka. Browne in reporting his proceeding to the Chief Commissioner, pointed out:

> Lengpunga's deposal will not only be a severe punishment to the man himself, but will effectually dispel any suspicion among the people that our action in not persisting in his surrender is due to weakness. In the size of his village and power generally, he was second only to Khalkama, while in influence and in the fear which

---

150 *Ibid.*, (No. 134) Statement made at Camp No. 12 outside Lengpunga's *punji*, 9 February 1890.

151 *Ibid.*

> he inspired, both among his own people and among other tribes, he appears to have been the leading chief in the country round. Jarak, on the other hand was a chief only in name.[152]

Browne decided to inflict lesser fine in guns on Zahrawka, as he was much inferior not only to Lianphunga but also to the other chiefs. As he had explained:

> His mother not being a Rani, his status as a chief is far below that of Lengpunga and the other legitimate sons of Sukpilal, and since the junction of his punji with Lengpunga's, his followers and property appeared to have been absorbed by the latter. His share of the village property, as stated to me by Lengpunga, consisted of only two metnas and 500 maunds of rice. It was no doubt a consideration of this difference in their position which led the Bengal Government to constitute Lengpunga the main offender, although Jarak was the first to suggest the raid, and was the actual leader of it.[153]

Browne hoped that by fixing a limit to the period of their sentence, which at one time he was inclined to make perpetual, and holding out hopes of re-instatement, they might be induced to accept their fate with equanimity. On the other hand if the punishment was permanent, he thought it probable that having nothing to hope for and no inducement to good behaviour they might directly or indirectly cause him more trouble in future. Browne also realised that as neither Lianphunga nor Zahrawka could have a village of his own, as per the sentence of their punishment, it would be difficult to punish them in case they misbehaved. He, therefore, made four of their brothers, a nephew, and five leading chiefs, as their sureties. He firmly believed that if sufficient pressure was brought to bear on them they would be able to keep the two offenders under restraint.

---

[152] *Ibid.*, (No.133). Browne to Secretary to the Chief Commissioner , Assam. 14 June 1890.

[153] *Ibid.*

In determining whether in addition a fine in cash or kind should be imposed, Browne took into consideration the loss which Lianphunga and Zahrawka had already suffered at British hands. Their villages had been burnt to the ground during the expedition of 1889-1890, and everything found in it had been destroyed. Under this circumstance Browne considered it, "undesirable to impose a further fine in marketable property, which would probably only be paid by other villages, already somewhat harassed by having for nine months of this year to support the large population comprising Lengpunga's following."[154] It was quite a different matter about guns. There were, Browne suspected, a large number of guns with the Lushais since only a few had been surrendered in the late expedition. He therefore insisted upon realising fines in guns.[155]

The Political Officer's announcement at the durbar, of the imposition of the revenue and the labour arrangement at once caused murmuring amongst the chiefs. It was the one issue that they deliberated at length and the result was not as conclusive as Browne might have liked.[156] In fact it was clear that the chiefs were not prepared to accept these impositions

---

[154] *Ibid.*

[155] *Ibid.*, He added "A fine in fire arms, too has a far greater effect than one in property, especially nowadays, when there is no opportunity of replacing them, as was the custom until recently, through Burma. Mr. Daly considered that if a fine of guns was imposed, ten would be a proper number, and Rai Hari Charan Sarma Bahadur, who for many years has annually made two or three trips in to these hills, and whom I recently consulted as to the number of guns the Lushai chiefs possess, was of opinion that, as a rule, a fine of guns should not be less than five or more than ten. I myself was inclined to confiscate a somewhat larger number than Mr. Daly recommended, but in view of the opinion expressed by him and the Rai Bahdur, and bearing in mind that six have already been taken, I thought a further fine of 15 guns would meet the requirement of the case. Of this number, ten were to be paid by Lengpunga and five by Jarak. Lengpunga and his brother agreed without demur to pay his share of the fine, but it was only after some delay and consultation that the chiefs undertook to be responsible for the five guns to be realised from Jarak, Khalkam alleging that the latter had only one or two guns of his own at most."

[156] FEAP. January 1892; Nos. 51-61, McCabe to Secretary to the Chief Commissioner of Assam, 11 June 1891.

and that there would be some opposition to them later on. Captain Browne concluded the *darbar* by telling the chiefs:

1. That in future there must be no more fighting anywhere among themselves; if attacked, they could defend themselves, but they were not to attack any village in retaliation; that fighting with villages in the south-west of Manipur territory which was going on for years was to be discontinued; likewise of these villages were attacked; they would be required to report to Browne to get their grievances redressed. He made it plain to the chiefs that fighting except in self-defence, would be severely punished.
2. That British subjects, sepoys, coolies, followers, traders etc., might be allowed to go about freely in the Lushai hills without the fear of molestation. If they experienced any annoyance from any British subjects entering their villages, they were to inform the Political Officer, who would take appropriate action.
3. Any British subjects guilty of committing an offence against a Lushai, the people and chiefs were on no account to take the law into their own hands, but to report to the Political Officer who would investigate and redress grievances.
4. As the people were reportedly anxious to have bazar at Changsil, this would be opened. If, however, the Lushais considered the prices of articles were charged by the traders to be too high, the Political Officer would fix the rates. But it was pointed out that reasonable price would be fixed, considering the long routes the traders would have to take and considering the reasonable profits for them in order to keep in bussiness.[157]

REACTION: WESTERN LUSHAIS UPRISING

During the months of July and August, Captain Browne began to implement his policy regarding revenue and labour.

---

[157] FEAP. October 1890; Nos. 126-144, Browne to Chief Commissioner, Assam, 14 June 1890.

Repeated attempts were made to effect the preliminary settlement of revenue and the procurement of the labour from the Lushai chiefs. Browne's persistence in this regard is surprising. Not only was he told by the Chief Commissioner not to insist upon payment of revenue if the chiefs were disinclined to pay but there was already evidence that the chiefs would resist collection. There was of course another reason for disaffection. Punishment inflicted on Lianphunga and Zahrawka, which deposed them from their chieftaincy even for the limited period of four years was considered harsh by the Lushais. Under Lushai custom no chief, however oppressive and disliked, could be removed: chiefs captured during raids were always ransomed and in no case were they removed from their position.

Smarting under the impositions made by the Political Officer and unhappy with the treatment of Lianphunga and Zahrawka the Lushai chiefs near Aizawl, all located west of the Tuirial and hence called the western Lushais made up their minds to resist. Kalkhama, who took a leading part in the anti-British movement, was later to explain what drove him to desperation:

> ...I took an oath of fealty with Browne saheb, and he told us that we would have to pay revenue. We did not like this, and subsequently when Leipung, a mantri of Liankunga, told me that he had heard the sahib was going to collect from the western Lushais first and then from the Eastern Lushais, and that we would not even be allowed liberty to hunt in the jungle, I lost my head and resolved to fight.[158]

Neither Lianphunga nor his sureties, it was becoming increasingly clear, had any intention to act upon the decisions arrived at the *darbar*. Lianphunga's *mantri* in fact bitterly complained when warned by Browne that the new village set

[158] FEAP. January 1891; Nos. 121-134, H.W.G.Cole, Commandant, Surma Valley Military Police, to the Inspector General of police, Assam. (Through the Political Officer, North Lushai Hills.) see Statement of Kalkhama.

up by Lianphunga in the meantime would be destroyed. Discontent was now fairly widespread among all the western chiefs.

Captain Browne was not totally unaware of the situation or that the chiefs had decided to save Lianphunga and Zahrawka from further punishment.[159] He knew at the same time that they were oblivious of the consequences of opposing the British. The situation now became complicated, and Captain Browne even briefly considered changing some of his orders. But he finally decided against it for he was convinced that the punishment of the guilty chiefs was more important than the settling of the country. That he had persisted in his demands for revenue and labour unwisely and in the face of mounting disaffection is brought out by Robert McCabe, who succeeded him, by a close examination of the latter's diaries:

> Visits were subsequently paid to Lalrhima, Thalien, Rankupa, and Sailenpui, and in each village signs of a determination to oppose at least passive resistance to all attempts either to collect revenue or supply labour were visible. The chiefs had fully made up their minds to prevent the creation of any precedent for the supply of coolies on any terms, and even declined to allow their men to act as Captain Browne's chaprasis. Finally we find that on the 25th. August, in the last diary submitted by Browne, Lalbura and Bungteya's mantri came in, and, speaking on behalf of all the eastern Lushai chiefs, absolutely declined to pay revenue.[160]

The western Lushai chiefs were mainly the descendants of Suakpuilala or their dependants*. All of them ruled separate villages independently of one another. But they were all related by blood and linked together in some sort of a confederacy.[161]

---

[159] FEAP. October 1890; Nos. 126-144, Demi Official letter from Browne to the Secretary to the Chief Commissioner of Assam.

[160] FEAP. January 1892; Nos. 51-61, McCabe, report to the Chief Commissioner of Assam. 11 June 1891.

*For the genealogy of the western Lushais see Appendix No. 1.

[161] FEAP. January 1891; Nos. 121- 144.

They consulted each other under the leadership of the eldest among them in all matters relating to warfare, boundary disputes, common welfare of the people, and so on. Kalkhama was the eldest among them and so the leader of the whole of the western Lushais.

Browne's action quickly brought them all together. On 1 September 1890 at Kalkhama's *zawlbuk,* where all of them met, they discussed the measures to be taken to expel the British from their hills. All of them agreed that they should strike first, as a statement from one Ruma makes clear:

> ... as war had to be declared, it was advisable that the initiative should be taken by the Lushais, as if the British government took them unawares a lot of property would be lost, while if the Lushais attacked the forts and killed the sepoys and coolie no more men would be sent up from Silchar.[162]

The responsibility of each chief were formally assigned: Kalkhama's men, 100 in number, with the same number of Thanghulha's, Liankunga's and ten of Tulera's were to watch Changsil; Lalhrima's men were to attack the coolies on the road, while the men of Thanruma and Sailianpuia were to keep a watch on Sairang.[163]

After their duties were assigned and arrangements completed each chief was asked to keep himself and his men in readiness to strike at any moment. It now remained to follow the movements of Captain Browne so as to determine when and where to attack. As Thanruma's village was nearest to Fort Aizawl, he was told to keep others informed about the movement of Captain Browne, and all others were to take every opportunity to obtain whatever information they could

---

[162] FSEP June 1891; Nos. 80-100, R.B.McCabe, Political Officer, North Lushai Hills, to the Secretary to the Chief Commissioner of Assam. See Statement of Ruma sent by Kalkhama on 18 October 1891.

[163] FEAP. January 1891; Nos. 121- 134, see Statement of Kalkhama.

procure for the success of their endeavours.[164] Thanruma was able to pass on vital information about Browne and the garrison at Aizawl on 8 September 1890.[165]

Thanruma on the same day met Captain Browne and to allay any suspicion showed outwardly friendliness and submissiveness.[166] The political officer appeared completely fooled by Thanruma, for even after his orderly, who saw number of armed Lushais proceeding towards Sairang, told him that "These Lushais mean mischief." Browne replied: "Your head is turned with fear".[167]

After meeting Thanruma Captain Browne proceeded for Sairang on his way to Shillong to discuss Lushai affairs with the Chief Commissioner. This news was immediately passed on to the other chiefs by Thanruma.[168] And by the time Browne left Sairang for Changsil all the Lushai chiefs and their men had been informed and taken their respective positions. In the early morning of 9 September 1890, Captain Browne left Sairang for Changsil with one clerk, two followers, twenty five coolies, one *dufadar*, and four military police constables. Meanwhile twenty three coolies and two constables were on their way from Aizawl for Sairang, and one *havildar*, six men, and four coolies were likewise returning to their headquarters at Aizawl.[169] All three parties left their respective posts about the same time.

The Lushais, made simultaneous attacks on these three parties. Their main aim was to cut off all the communication

---

[164] FSEP. June 1891; Nos. 80-100, McCabe to Secretary to the Chief Commissioner of Assam, 19 January 1891.

[165] *Ibid.*

[166] *Ibid.*

[167] *Ibid.*, Liankhama's *mantri*, who had acted as political chaprasi for the western Lushais and who was accused of bęing in the pay of the British was warned to keep silent of these developments. The man abruptly left Aizawl on the 28 August 1890 but he failed to inform the Political Officer that an uprising was being fermented.

[168] *Ibid.*

[169] *Ibid.*

between each party, so as not to be able to assist one another. Captain Browne was attacked by the men of Liankunga, Kalkhama, Tulera, and Khama. They were actually assigned to attack Changsil and guard the first three miles of the road toward Aizawl. Liankunga's men fired the first shot at Browne, but it was one Saithawma's gun that hit Browne. Browne along with one servant retreated and managed to reach Changsil post but soon succumbed to his injuries.[170] While the attacks upon Browne and the Changsil fort were going on, Aizawl was assailed in strength.[171] All of them were apparently taken by surprise. Virtually no resistance was offered by the police. The coolies naturally threw down their loads and bolted when they saw the police instead of fighting back the Lushais were running off.[172] The loss of the British in this attack was one European Officer, a clerk, five coolies, five traders and two *manjis* or boatmen.

The Lushais cut off all communication between the Lushai Hills and Cachar and between Aizawl and Changsil. Relief was therefore not immediately forthcoming either from the plains or between one another in the hills. Nonetheless Aizawl and Changsil Forts were however held in the face of repeated attacks between 9 and 28 September.[173] The men who were closely invested inside their stockades would have run short of supplies which could last no longer than a couple of days longer. They had suffered great hardship from rain and cold, and constant duty to hold back the investors. Moreover, they did not even possess warm clothes and enough drinking water, a sad commentary on the state of affairs in the Lushai Hills.

Lieutenant Harry Cole, commanding the Changsil Fort, somehow managed to send a boat overnight to Jhalnachera

---

170 FEAP. January 1891; Nos.121-134, see Statement of Tulera.

171 *Ibid.*

172 FEAP. December 1890; Nos.253-271, Demi Official from McCabe to the Quinton.

173 FEAP. January 1892; Nos. 51-61.

at the foot hills to inform the Chief Commissioner of Assam about the death of Captain Browne and general outbreak. At the same time he managed to open communication with Aizawl Fort through heliographic signalling.[174]

On the receipt of Cole's information, a relief column of frontier police under Lieutenant Tytler was immediately rushed from Silchar on 12 September 1890. This was joined by a party of Gurkhas under Lieutenant Swinton who assumed the command of the operations. The combined column moved up the Tlawng river in a convoy of boats. While approaching Liankunga's *jhums*, the flanking party that moved along the banks of the river had to called in to assist in dragging the boats through the strong current. Taking advantage of this the Lushais attacked the relieving force. Though repulsed, the British lost Swinton, a *sepoy*, and a *manji* killed and one *manji* and three *sepoys* wounded.[175] Nevertheless, the column pulled through to reach Changsil and relieved the Fort on 28 September. Two days later a further reinforcement of 200 men of the 40th Bengal Infantry, arrived at Changsil and Lieutenant Watson who commanded the detachment took over the charge of the mixed military and police force. On 4 October 1890, they were able to relief Aizawl.[176]

McCABE'S OPERATIONS

Robert McCabe, who had distinguished himself in suppressing the Ao Nagas, was appointed to succeed Captain Browne on 26 September 1890.[177] Without loosing time McCabe proceeded from Shillong to Cachar. On 1 October 1890, accompanied by Captain Williamson, and twenty four frontier police McCabe left Cachar's frontier out post Jhalnacherra and reached Changsil, covering a distance in

---

[174] *Ibid.*
[175] *Ibid.*
[176] *Ibid.*
[177] FEAP. January 1892; Nos. 51-61.

four days which a convoy normally took eleven.[178] On the following evening Hrangkhupa's men sneaking through the *jhum* fields on both banks of the river Tlawng attempted to harass the troops.

Hrangkhupa's village situated on a ridge some 3500 feet high a few miles to the west of Changsil commanded the river communications. His subjugation or to "squash him," as McCabe put it, became of immediate importance to the security of the boatmen. On the midnight of 6 October McCabe marched out against the village with a hundred men. At about one in the morning he reached the ridge. McCabe's account of the action shows the nature of warfare that was to take in these hills:

> We marched through dense bamboo jungle till 5:30 a.m., then passed through some jhums and at 7 a.m. entered tree jungle. All of a sudden, we came on a stockade, and a bullet whizzed through a sepoy's cap. No damage was done except government property, and, on firing a volley, we found the stockade abandoned. We then proceeded cautiously, and came on a second stockade. This was taken in flank, and no opposition experienced. As we neared the village, the Lushai paid us all attention, snipping at us as we crossed the jhums. Finally, the village was rushed at 9:30 a.m. and abandoned by the enemy.[179]

"We utterly effaced the village" McCabe concluded. Thereafter he retired to Changsil, carrying with him fowl and goats, the whole operation taking nineteen hours of a constant gruelling march.

Meanwhile on the 2 October a relief column of 250 Frontier Police and 50 men from the 40th Bengal Infantry moved out from the Changsil fort to Aizawl under Lieutenant Cole. The force halted in Sairang on the following day, where a stockade

---

[178] FEAP. December 1890; Nos.235-271, McCabe to Quinton, 20 October 1890.
[179] *Ibid*.

was built and fifty men posted there for security of the line of communications. The following day Cole and the 250 men moved out to Thanruma's village, which commanded the Aizawl-Changsil road and intercepted each communications. Though slight opposition was met the village was destroyed and two cows and three calves were carried away. With communication with Changsil thus secured Cole and his column moved up to Aizawl.[180]

For the next week McCabe devoted his energies towards strengthening his position in the hills. Reinforcement had to be brought in, supplies collected and coolie transport reorganised. There was a good deal of sickness among the troops and coolies, some of whom had to be sent back to Cachar. Reliable information of what led to the uprising was still not forthcoming. "I would like to make peace with one of the least guilty village," he wrote to the Chief Commissioner, "so as to open negotiations with the Lushais generally, and obtain information of what was going on." Still, some of the smaller village had to be taught that "the cockerel cannot imitate the cock with impunity, and a good lesson now will prevent them from following Khalkam and Lengpunga's lead in the future." He would have liked to attack Kalkhama without a delay but the road was difficult and he was short of coolies. "We have only 50 real coolies available, and these are nothing for 300 sepoys for ten days" he had added. A convoy of forty-eight boats arrived on 13 October bringing in eighty coolies and about a thousand maunds of supplies. This made it possible to organise a small column to attack and destroy within the next week the villages of Hmingthanga, Lalsavuta and Liankunga, all within a radius of five miles from Changsil

[180] *Ibid.*, In Changsil McCabe found many men suffering from scurvy. Lime-juice and hospital comforts were no-existent, and fresh meat and vegetable were not obtainable.

on the road to Aizawl.[181] From Aizawl Lieutenant Cole moved again to Lalhrima's village and destroyed it on the 14th [182]

By the close of October it was clear that all the western Lushais were involved in the uprising. The month's operations had already frightened the Lushais. But the chiefs had not come in, and McCabe was convinced that they may be only "delaying with us so as to gain time to get in their crops." He therefore decided to "settle the business once for all" to save the expense and trouble of fresh expeditions. He sought to enforce the unconditional surrender of the chiefs through the establishment of posts within their villages and by harrying them as much as possible and destroying their crops. This policy was carried out by McCabe in the first week of

---

[181] *Ibid*. McCabe to Quinton, 14 October and 19 October. The villages were not very large; Mingthang, 150 houses; Lalsavuta, 180 houses and Lenkhai, 80 houses.

[182] *Ibid*. Cole to Inspector General of Police, 15 October 1890. Cole thus described his action: "A start was made from Aijal at 5 a.m. The road even for a Lushai path is very bad one, and, owing to the recent rain, progress was difficult and slow. I reached a high peak 5000 feet high, and about four miles from Aijal, at 8 am. And found two stockades which the Lushais had erected at its summit, also a white shirt of Captain Browne's, which they had placed on a long bamboo as a flag. Both stockades were unoccupied. From this point to the village the road is fairly good. On the slope of a hill about one mile from the village, the Lushais, had cleared the jungle for some distance on each side of the road, so as to be able to see any force that might approach. From here to the village we found five more small stockades erected at different places along the road. All these were unoccupied, but a number of Lushais collected in the village, and tried to make a stand. There was a stockade round the portion of the village which commanded the village. On approaching the village I extended the men into the jungle on each side of the road, sending one section to attack from the right flank along the road from Lalbura's, with an advanced rapidly, firing volleys. The Lushais all bolted as the order was given to charge. It is impossible to estimate the number the Lushai lost, but two men were seen to fall to our volleys, and a gun, smashed in two by a bullet was found in the village, while wounded men were heard shouting for assistance from the jungle. I occupied the village at 10 a.m. A large quantity of goats and fowls were found, which we brought away with us." The Lushais evidently we taken by surprise. Women were observed performing their daily chores. Some carrying loads on their back, while other were fetching water from a stream below the village. In the chief's house was found a large quantity of newly-cooked rice, still hot.

November. A telegram to the Chief Commissioner shows what he had done:

> On 7th and 8th made simultaneous attacks on Menthang, Rankupa, Lalrhima and Thanruma. The Minthang-Rankupa party consisted of 80 men of 40th Regiment and 18 Frontier Police, with Lieutenant Watson, Mr. Shuttleworth, and Dr. Whitchurch. They destroyed Rankupa and Minthang's jhums and jhum-houses, also six new settlements erected in the last fortnight, besides considerable household property.[183]

McCabe further added:

> The Lalrhima party, with 100 frontier police under Lieutenant Tytler and Dr. Melville, destroyed Lalrhima's jhums and jhum-houses, and brought away goats and fowls. They were fired on continuously all night, and we had two men slightly wounded. The main body of 150 men under Mr. McCabe and Lieutenant Cole proceeded to Thanruma's old site, and sent off parties to destroy jhums, and Thanruma's new village ten miles to the south- west, burnt 160 new houses, killed five I ushais, and brought in 22 goats and 300 fowl.[184]

Thus McCabe resorted to extreme violence, burning houses and dispersing the inhabitants, slaughtering livestock and destroying grain, to enforce obedience to British authority.

One village after another in the western hills was overrun by the British. But Kalkhama, held on defiantly: he constructed formidable stockades on the approach to his village from Changsil, the direction from which he anticipated McCabe's attack.[185] Unfortunately for him though a flying column advanced from this route as a diversion, the main force came from the Aizawl route which lay behind the village and simultaneously charged Kalkhama's defences.[186] Caught by surprise and realising that they were attacked from the

183 *Ibid.*, No. 262, McCabe to Chief Commissioner, 14 November.

184 *Ibid.*

185 *Ibid.*

186 *Ibid.*

rear as well as from the front the Lushais gave up their resistance and fled with the loss of ten men. Search parties were sent out to the jungle to collect concealed grain and domestic materials which were removed before the attack and to hunt down the Lushais. The village was destroyed and all houses burnt, so too were *jhums* and *jhum* houses. In the end a large number of personal effects, kitchen utensils, *mithuns*, goats, and fowl were collected.[187]

SURRENDER OF CHIEFS

Extensive destruction of villages compelled the Lushais to seek shelter in the jungle where they were still pursued by British troops, and even their temporary shelters with all their belongings destroyed. In these circumstances Kalkhama realised the futility of further opposing the British. He accordingly sent his *mantri* and that of Lianphunga to effect a settlement and to plead for the relief of their miserable condition.[188] McCabe was not prepared for any settlement other than the unconditional surrender of the chiefs. Not all the Lushai chiefs were ready to surrender, and it was important to ensure this by the surrender of Kalkhama who was undoubtedly the leading one among them.

Consequently McCabe sent a political *chaprasi*, Changkunga's brother, one 'Bhaya', to convince Kalkhama about surrendering unconditionally. Bhaya* met Kalkhama who expressed his willingness to come in on McCabe's terms. Following this Kalkhama surrendered to McCabe on the 22 November 1890.[189]

McCabe was fully aware that Kalkhama was the acknowledged successor of Suakpuilala and the head of the western Lushais. He thus described the situation:

---

[187] *Ibid.*

[188] *Ibid.*

* Probably 'Vaia", there is no Lushai name 'Bhaya'.

[189] FEAP. January 1892; Nos. 51-61.

> Sukpilal died in 1880, and his youngest son Lalrhima became chief of the head Western Lushai village, which was at that time situated near the present site of Aijal. Lalrhima was a child of seven years of age, entirely in the hands of his mother, the junior Rani, and in those of an old mantri named Lalcherra. This type of ruling authority was not calculated to attract Lushai subjects, and we find that many families preferred to place themselves under the more powerful protection of Khalkam, Lengpunga, and Sailenpui. Gradually the main authority over the villages in the east of Dhaleswari centred in Khalkam, while that over the villages on the west bank was assumed by Sailenpui, Lengpung maintaining a kind of independent position.[190]

McCabe therefore sought to use Kalkhama's influence in inducing the remaining Western Lushai chiefs to surrender. He told Kalkhama that as soon as he, Kalkhama, was able to call in all his brothers, orders would be passed on his case and the others involved in the uprising.[191] Eager for an early settlement of his fate Kalkhama despatched messengers to Tulera, Khama, Liankunga, Thanghulha and others to lay down their arms. Of these, Tulera and Khama surrendered to McCabe within two days.

Thanghulha remained adamant and held aloof. His *jhums* were at once burnt, and one of his men was killed. This was followed by an ultimatum that if he did not immediately surrender his village would be burnt. Thanghulha ignored the ultimatum and instead he evacuated the village. It was promptly destroyed and set on fire. This action had a powerful effect upon Liankunga who fearing that his village would also be burnt quickly surrendered.[192]

Once again, McCabe used Kalkhama to get messages across to various western Lushai chiefs. As a result of his intercession, chiefs like Lalhrima, Sailianpuia, and

---

[190] *Ibid.*
[191] *Ibid.*
[192] *Ibid.*

Thanghulha tendered their submission to McCabe between the 2nd and 5th December.[193] Lalhrima gave up Captain Browne's gun, rifle, two revolvers, two watches, a pair of binoculars and an anteriod,[194] all in fair condition. He also surrendered the guns confiscated by Captain Browne.[195]

Lianphunga held out a little longer.[196] But when Lieutenant Cole with hundred men went out after him, he too surrendered on the 9th.[197] By the close of the year with the exception of Thanruma who had taken refuge in the extreme southern border of the North Lushai Hills district all the Western Lushai chiefs had laid down their arms.[198]

DEPORTATION : KALKHAMA, LIANPHUNGA AND THANGHULHA

Following their submission, McCabe placed the chiefs on trial. A number of them, admitting their guilt, had brought in some guns at the time of their submission. They were made to surrender more as a fine. Of these Liankunga, Lalhrima, Hrangkhupa, Lalluaia and Lalsavuta were considered too young to be held personally responsible for the late rising; whereas Tulera, Thawmpawnga, Hmingthanga and Thatliana, were completely under the control of Kalkhama and Sailianpuia.[199]

·McCabe therefore realised that he had only to deal with Kalkhama, Lianphunga, Sailianpuia, and Thanghulha, as

---

193 *Ibid.*

194 *Ibid.*

195 *Ibid.*

196 *Ibid.*

197 *Ibid.*

198 *Ibid.*

199 Reid, *op,cit.*, p. 26.

·McCabe described the necessity of the harsh punishment to be inflicted to the hostile Chiefs thus :

"The outbreak of last September proved clearly that the punishment inflicted by the Expeditionary force in 1889-90 was quite inadequate. It was no use firing at the tiger with a pea shooter, or attempting to extract a shark's tooth with a kid glove as protection; and it is equally useless to expect to impress Lushais with a sense of our power by merely burning their villages, which cost them nothing

independent agents controlling the actions of the whole of the Western Lushais. It was these four that required stern action to be taken against them. But equally important was the need, as McCabe had also realised, of at least one leading or powerful chief to collect the disintegrated unit of the Western Lushai families. He therefore left Sailianpuia out in order to utilise his power in controlling the people.[200] As regards the uprising Sailianpuia had relatively a better record than the other four, though his men had taken part in the looting of the Changsil Bazar. McCabe referring to Browne's note on the Lushais, observed:

> In Captain Browne's note on the Lushais referring to this chief, I find it recorded that there is probably no other Chief who has universally given such satisfaction in his dealings with us. And, as far as I could ascertain, he took a very minor part in the outbreak. His men undoubtedly joined in the looting of the bazar, but it would be as difficult to hold Lushais in check with such an opportunity before them as to keep flies from honey.[201]

On the other hand, Kalkhama, Lianphunga, and Thanghulha were undoubtedly actively involved in the rising. Kalkhama was certainly its leading spirit and all the details of the plan for offensive measures took shape under his guidance in his *zawlbuk*. Lianphunga was a man of equally strong personal character, able, energetic, and quite capable of holding his own in any tribal community. He was immensely trusted by his father over all his other brothers. Indeed as a defence against any attack from the south, from the Pawis and Lakhers, his village was located in the southern end of their territory. For his involvement in the Chengri valley raid he was punished with the destruction of his village and with a fine of ten guns. He had rebuilt his village soon enough and

---

but a little labour to rebuilt. They must be made to understand that a paramount authority which brooks no disobedience to its order exists, that it is essential they should open up their country with roads, and pay revenue as a sign of submission."

[200] FEAP. January 1892; Nos. 51-61.

[201] *Ibid.*

joined the outbreak of September 1890.[202] Thanghulha was of a turbulent nature and he was the one chief who alone led his men in the attack, for seven days, on Aizawl and Changsil forts and on Captain Browne's party.[203] He was the son of Suakpuilala's brother Runphunga, and Kalkhama's camp follower. He never bore any good will towards the British.

In view of these facts, McCabe decided to deport Kalkhama, Lianphunga, and Thanghulha. The three were sent under escort to Tezpur on the 13 January 1891.[204] The other chiefs were let off after they agreed to the demands to surrender guns and to supply the required number of punitive labour.[205]

It was clear to McCabe that the rising was a large and well organised affair and only an informed body of Lushai chiefs could possibly have made so many simultaneous attacks over a long line of twenty miles in length.[206] It was equally clear that planning was carefully done under a strong single leadership. To ensure that such extensive uprisings did not recur and that no single chief became all too powerful and assumed leadership, McCabe proposed three important measures to be adopted: firstly, to see the western villages rebuilt and get together all the scattered families, secondly, to give the Lushai an opportunity to prepare their *jhums* for cultivations that year, and finally, to re-establish responsible authority in every village.[207].

During the entire episode, the eastern Lushai chiefs were not actively involved, although Pawibawia and Lalburha later admitted that their men were involved in the looting of Changsil *bazar* and where the former's men was killed.

---

[202] *Ibid.*

[203] FSEP. June 1891; Nos. 80-100.

[204] FEAP. January 1892; Nos.51-61. They were taken to Hajaribarg jail where Kalkhama and Lianphunga later committed suicide. Thanghulha was released not long afterwards.

[205] *Ibid.*

[206] FSEP. June 1892; Nos.80-100.

[207] *Ibid.*

3

RESISTANCE IN THE EASTERN HILLS

It is not without reason that the eastern Lushais kept themselves aloof from the struggle of their western kinsmen against the imposition of British rule. The feuds between them which existed even before the expedition of 1871-72 had intensified after the death in 1880 of Suakpuilala, the 'grand old man' of the Lushai hills. Apparently the eastern chiefs saw the destruction of their rival's power at the hands of the British as a guarantee to their own survival as powerful and independent people. They hardly anticipated that they too would soon be required to pay revenue and supply labour. The September uprising was therefore less extensive than it otherwise could have been. Robert McCabe clearly understood this situation:

> As far as I could ascertain, the Eastern Lushai chiefs gave no concerted aid to the Western Lushais in the rising of September 1890. A few of their young warriors joined in the investment of Aijal and Changsil, and shelter was afforded to refugees from Kalkhama and Thanruma, but apparently no organised assistance was rendered. We have not far to seek the motives that prompted the Eastern Lushais in this line of action. They had for many years carried on incessant feuds with their Western neighbours, and could view with equanimity the destruction of their formidable and rival power created

> by Sukpilal, while, on the other hand, they did not anticipate that the British Government would exercise any control over their actions or attempt to collect tribute from them.[208]

McCabe rightly believed that should a struggle with the eastern chiefs ensue the latter could expect no assistance from west of the Tuirial. On the other hand the relations between the eastern chiefs were far from friendly. The group consisted, it may be recalled, of the descendants of Vanhnuailiana, Lalphunga and Vuta.* In so far as power was concerned, the villages of Pawibawia, Lalburha, Buangtheuva and Liankhama formed one group, while those of Vuta's descendants formed the other. McCabe did not expect any combination between these two groups:

> No combination may be anticipated between these two groups, but it is absolutely certain that all the members of the first group must be looked upon as one unit guided and moved by a single policy. Lenkham and Bungteya Raja died within the past two months, and Pawibawia (village) is ruled by a minor so that the sole remaining powerful chief is Lalbura. I proposed to make a camp in this latter chief's village, and use it as a base from which to collect revenue.[209]

Lalburha was recognised as the most powerful. He controlled the villages of Buangtheuva, Chinhleia, Liankhama, Kaihranga and Nikuala, which comprised more or less 2135 houses in addition to his own village of 474 houses. At the same time, he had great influence over the Pawibawi group of villages ruled by the latter's minor sons, namely, Vanphunga, Thangkama and Lalruaia consisting of 200, 722 and 60 houses respectively. Also included in his sphere of influence, were the villages of Pawibawia's mother, Saitual containing 180 houses, Lalhleia village of 100 houses and

---

[208] FEAP. December 1892; Nos. 42-46, McCabe to Chief Commissioner, 23 July 1892. Report on the outbreak of the Eastern Lushai Tribe.

*See Appendix II. For genealogy of the Eastern Lushai chiefs.

[209] *Ibid.*

Pawibawia's half brother Lalruma's village of approximately the same size.

THE PROMENADE : REVENUE AND LABOUR IMPOSED

Not anticipating any serious opposition from any quarter, McCabe decided to undertake a military tour, or what was then called a promenade, of the eastern villages in early 1891. It was not to be a punitive measure, but McCabe was clear in his mind that if any resistance was offered to him he would use force to crush it. The objects of the promenade, as McCabe himself indicated, were two: first, to visit all the villages east of the Tuirial, locate their sites and make the acquaintance of the chiefs; secondly, to inform these chiefs that they were now under the British Government, and would henceforth have to pay revenue.[210]

He took with him a force consisting of 400 men of 43rd. Gurkha Light Infantry and 20 Military Police* commanded by Colonel Evans, and an initial contingent of 320 coolies drawn from the Public Works Department. As in all hill or frontier campaigns commissariat and transport arrangements proved most difficult to solve:

> The problem we had to face (McCabe said in his report) was how to obtain transport. On 320 coolies fast sufficed to move the baggage, equipment, and ammunition of the force and no carriage was available for the transport of 1200 maunds of rations. It is true that Lalbura and Bungteya had agreed to assist with coolies and rice, but no absolute reliance could be placed in their promises. In the Naga Hills I have often gone for long promenades, obtaining supplies and coolies from village and village, and I saw no reason why the same plan should equally successful in Lushai land.[211]

---

[210] FEAP. January 1892; Nos. 51-61, McCabe's Report on the North Lushai Hills for the year 1890-1891.

*In late 1890 the Frontier Police in Assam was reorganised as the Assam Military Police and territorially organised into four Battalions.

[211] *Ibid.*

McCabe therefore decided to establish staging posts as he proceeded eastwards. The first was on the Tuirial. Lalburha and Buangtheuva were made to supply men to clear the site for the camp which was connected by a new and direct road to Aizawl. The first half of the road was made suitable for mule transport and by 23 January all supplies were brought in. Here the *mantris* of Lalburha did

> all in their power to throw impediments in our way and dissuade us from entering their country, stating that the Rajas were still consulting and that no more than ten coolies could be obtained from each village with very little rice.[212]

McCabe was not surprised at this opposition, as he had understood that:

> If Lalburha had received us with open arms he would have incurred the enmity of the other chiefs and the advise of having introduced a so called alien power into their country.[213]

McCabe brushed aside these objections and told the *mantri* that he would march into Lalburha's village the next morning. The *mantris* made one more attempt to stop McCabe saying that if he insisted upon proceeding the chief and his people would desert the village.[214] McCabe thereupon tried to allay his fear telling him that he only intended to see the chiefs to settle the question of revenue, administrative matters and fix the future relations between them.

The next day, McCabe marched towards Lalburha's village, leaving 100 men behind to guard the Tuirial camp. On his way, he met sixty Lushai coolies sent by Lalburha to carry the baggage and supplies. They too protested against the entry of the force into their village. The protest was ignored and McCabe entered the village without serious opposition, and a central position in it was occupied by the

---

[212] *Ibid.*
[213] *Ibid.*
[214] *Ibid.*

troops.[215] As the *mantri* said he would, Lalburha along with some of his villagers had already left the village. In spite of this seven hundred coolies including those obtained from Pawibawia's village were supplied. This made it possible for McCabe to establish a camp in Lalburha's village.

McCabe demanded to see Lalburha personally. On the 27th the chief sent in messages that he was prepared to meet McCabe, but outside the village and provided that the latter came without an escort. McCabe had no reason to expect treachery and so along with Colonel Evans went out to meet the chief, whom he found had come with a strong bodyguard.[216] After some parleys Lalburha agreed to the payment of revenue at one rupee per house, to supply rice at two rupees per maund, calculated at 10 seers per house; and supply coolies on demand. Thereafter the old chief made a "hurried and undignified" departure into the adjacent jungle.

Two days later McCabe marched to Pawibawia's village with 200 *sepoys*, and encamped at Lalruaia's village Lailak.[217] He was received in a hospitable manner, provided with rice, fowl, etc. On the following day a meeting was held at Pawibawia's *zawlbuk*, attended by chiefs Lalruaia, Thangkama, Lalhleia, and Vanphunga, with *mantris* of some young chiefs.[218] McCabe extracted from them the same terms that he had made with Lalburha. On their part the chiefs expressed a desire for the re-establishment of the Tuirial *bazar*, and also asked for assistance in checking the raids of the Paite, who inhabited the country on the Lushai-Manipur frontier, to both of which McCabe agreed. As a token of agreement, he was supplied with hundred maunds of rice, *gratis*, which was delivered at Lalburha's village. In return McCabe promised

215 *Ibid*.

216 *Ibid*.

217 Lalchhanhima Chhangte and Biaktluanga, *Mizo Lal Ropuite*, (Aizawl, 1996). p. 162.

218 FEAP. January 1892; Nos. 51-61, McCabe's report on Eastern Lushai Promenade.

not to enforce the revenue and labour demands till November and December of that year that is, until the close of the harvesting season.[219]

From his base at Lalburha's village, McCabe next moved to Buangtheuva's village with 300 sepoys on the 3 February 1891, where he was received cordially.[220] Accommodation for the whole force had been arranged and as earlier directed 100 maunds of rice was collected for the troops, with pigs, goats, etc. The chief, Buangtheuva, confidently came out to meet McCabe, and spoke boldly without consulting any of his *mantris*.[221] When McCabe inquired of his complicity in the Chengri valley raid, he admitted freely that on Lianphunga's persuasion he had agreed to join the raiding party. But he turned back as he was informed by his brother Lalburha and by Liankhama that the village to be attacked was under the protection of the Maharani, and that raiding it would be an offence to the British.[222] McCabe could find no reliable reason to punish him and so only imposed revenue and supply of rice and labour.

Meanwhile, in order to simplify his task McCabe employed Khamliana, a former chief among the Haulawng group, but who had been driven away from them, and had taken refuge with his brother-in-law Sailianpuia. Tired of his dependent position, he wished to start a village of his own near Aizawl.* His family and dependants were in Lalburha's hands and without McCabe's help he had no hopes for their release. McCabe therefore employed him as a messenger to approach different Lushai chiefs, in return he promised to help him start the new village. Through him McCabe obtained

[219] *Ibid*.
[220] *Ibid*.
[221] *Ibid*.
[222] *Ibid*.

*Khamliana accompanied by McCabe throughout the promenade, "strict faith was kept on both side" wrote McCabe, and "he served me admirably, and I obtained the release of his relatives." These amounting to forty houses, were settled near Thanruma village.

information regarding Lalruma of Kawlkulh and Liankhama of Zawlnghak that he would be received in a friendly manner. He was also able to ascertain from Khamliana the route to be taken and the place for camping.[223]

On the 6th, McCabe proceeded to Lalruma's strongly stockaded village and was friendly received by the chief and the villagers. The terms for revenue and coolies were accepted and Lalruma supplied a hundred coolies, rice and vegetables.[224] After two days, McCabe and his force visited Liankhama's village, Zawlnghak, containing 852 houses. It was the largest among the eastern Lushai hills. A camp had already been constructed for the troops at a distance about 200 yards below the village. Here too he was received in a friendly manner and supplies were freely brought in to the camp. The chief, however, showed no generosity, but was well behaved towards the troops. After consenting to the terms imposed, he sought refuge in his own house, and McCabe thereafter returned to Buangtheuva.[225]

Three days later, McCabe marched to Lungliana's village, consisting of 225 houses. The chief made his disapproval of the visit known by retiring to the jungle. McCabe then called in his brother Kairuma, the most influential among Vuta's sons, and told him of the terms of revenue and labour supplies. This was accepted after which McCabe and his force returned to Lalburha's camp on 3 March 1891. Two days later he was back in Aizawl.[226]

Thus imposition of revenue and demands for labour was the most important aspect of McCabe's eastern Lushai promenade. It was too early, as McCabe himself recognised,

---

223 *Ibid.*

224 *Ibid.*, Lalruma as well reported Paite-Kukies ambuscaded on the bank of Tuivawl, which probably compelled to stockade his village very strongly and maintain strict guard. This report further confirmed the previous report given by Buangtheuva.

225 *Ibid.*

226 *Ibid.*

to assess what effect it had on the eastern Lushais, or whether house tax would be paid without demur after the next harvest. But he had no doubt that it was not totally without results:

1) We have increased our topographical knowledge of these hills, and have been able to locate the position of the villages which had not been previously fixed by survey.
2) The routes, camping grounds, and strength of the eastern Tuirial Lushais have been noted, and the relationship of the different chiefs to one another ascertained.
3) Relations have been entered into with the eastern Lushai chiefs, who all agreed to pay house-tax and supply rice and labour.
4) We have for the first time in the history of these hills demonstrated the fact that a force can march from village to village and rely upon obtaining Lushai coolies and supplies of rice and vegetables.[227]

SECOND PROMENADE

Between 23 January and 13 February 1892, McCabe once again marched through the eastern Lushai country. During this tour he intended to realise the house tax, demand labour, rice and vegetables. He met no immense difficulties in realising the house tax in cash from the villages that were visited, and coolies, rice and vegetables were freely supplied. However, the attitude of Lalburha and Buangtheuva changed even though they had accepted all terms put forth before them during McCabe's first visit.[228] Accordingly, McCabe fined Buangtheuva six guns for refusing to supply a few coolies to the escort that brought the rations from Aizawl to Lalburha's village. Lalburha too gave out that he could not force his men to do cooly work for if he did they would leave or *pem* to others

---

[227] *Ibid.*
[228] FEAP. December 1892; Nos. 42-46, Report.

village beyond his control. McCabe was not at all convinced with this pretext as he knew both Pawibawia's and Lalburha's men had for sometime past sought employment as coolies in carrying stores from Changsil to Aizawl.[229]

> a proof (said McCabe) that even from a Lushai standard the work was well remunerated. The chiefs objected to supply labour, as they themselves derived no benefit from it, and their subjects would naturally avail themselves of the excuse that they could not be expected to work for two masters... In fact, we may safely assert that the chiefs thought it derogatory to their position that their subjects should work for alien authorities.[230]

Lalburha, the most influential eastern Lushai chief, not surprisingly remained hostile to the British and now even denied having agreed to payment of revenue and supply of labour. Accustomed to a reputation as a powerful man in the eyes of his people and neighbouring chiefs he was not prepared to surrender this position before the British. He propagated the idea, not only among the eastern Lushai people but also among the western and southern Lushais that the motives of the British were suspect. He sent his *mantris* in February 1892 to Kairuma asking him to refuse payment of the house tax and join in a general rising with the object of driving out the British force from the Lushai hills.[231] However, Kairuma did not respond on the plea that his village could easily be attacked from both Aizawl or Lunglei and so he intended to comply with all orders of the British government.[232] Failure to enlist Kairuma's support not withstanding, Lalburha sought other allies and sent his men to draw the sympathy of all the Lushai chiefs between Aizawl and Vansanga's village

---

229 *Ibid.*

230 *Ibid.*

231 FEAP. December 1892; Nos. 42-46. McCabe to Secretary to the Chief Commissioner of Assam, 23 July 1892.

232 *Ibid.*

in the Haulawng country.[233] He assiduously disseminated the idea that McCabe's intention in asking for coolies was to confine them at Aizawl until he obtained possession of all guns in Lushai hands.[234]

McCabe's own reading of the situation was that Lalburha having lost his former prestige as a result of the recent promenade he was only provoking the other chiefs to join in an uprising to regain his lost power.[235] McCabe therefore continued with his attempts at enforcing revenue and labour demands upon the eastern Lushais. His idea was to realise these from each village, but first Lalburha had to be dealt with. On the 12 February 1892, on his return from the eastern Tuirial promenade, McCabe ordered that chief to supply 100 coolies to military movements between Changsil and Aizawl.[236] Lalburha promised that the coolies would be supplied within seven days. However, he had not the slightest intention of fulfilling his promise. Instead he sent emissaries to the western Lushai chiefs urging them to join in organising a general rising.[237]

ACTION AT SESAWNG: McCABE'S PLANS

By the middle of February 1892 many of the eastern Lushai chiefs had already inclined to submit to the British. On the 19th of that month, Pawibawia paid up the house tax amounting to Rs.700/- at Aizawl. McCabe told Lalburha to comply with his orders and that if he failed to do so, his village would be punitively punished. Lalburha promised his coolies at Aizawl on 26 February.[238] But they failed to appear on that date. McCabe decided to enforce his orders and two days later proceeded against Lalburha. His intenston was to force the

---

233 *Ibid.*
234 *Ibid.*
235 *Ibid.*
236 *Ibid.*
237 *Ibid.*
238 *Ibid.*

other chiefs, particularly the supposedly more powerful ones to submit to his orders. As he explained to the Chief Commissioner:

> We cannot assume that we have obtained thorough control until requisitions for labour meet with an immediate compliance; and, without this assistance the work of opening out communication would be delayed, and the civilisation of the country indefinitely postponed.[239]

He did not expect any serious opposition, confident that the sight of a moderate force would ensure immediate compliance. His escort was accordingly small, of 100 rifles, with Lieutenant Tytler, Assistant Commandant, Surgeon Captain Melville and two native officers from the Military Police and eighty-seven coolies. Each sepoy carried 70 rounds of ammunition and besides this, two spare boxes of ammunition and four days supplies were carried along with them.[240]

McCabe reached the Tuirial Camp the same day. Here the first resistance was met when Lalburha's men burnt the camp.[241] That a spirit of resistance was in the air is borne out by *dakwallas* declining to carry *dak* from Changsil, and their throwing off their loads and bolting to their villages at the news of the attack upon the camp.[242] On the 29th morning McCabe proceeded towards Sesawng, Lalburha's strong and large village. The villagers were apparently friendly: the usual presents of sugar cane and rice beer were offered. Coolies were promised within fourteen or fifteen days after *jhumming* operation had ceased. But ominously, the chief left the village and took to the adjacent jungle. McCabe occupied the *zawlbuk* for accommodation of his escort and the coolies.[243]

---

239 *Ibid.*

240 *Ibid.*

241 Sir Robert Reid, *History of the Frontier Areas Bordering on Assam*, (Shillong : 1942, reprint, 1978), p. 30.

242 FEAP. December 1892; Nos. 42-46. McCabe to Secretary to the Chief Commissioner of Assam, 23 July 1892.

243 *Ibid.*

The villagers too were seen busily engaged in removing their properties to the jungles and ravines nearby. Seeing all this McCabe informed the *mantris* that Lalburha must come in and meet him. It was not necessary for the people to remove their properties to the jungle, he told them, and that it would be foolishness on their part to be engaged in it. For he warned the *mantris* that his men were under orders to fire upon any Lushai armed with guns. The *mantris* still insisted that their chief would come and meet him, but took no steps to stop the exodus of the people.[244]

McCabe was further surprised when he saw considerable number of Pawis, who roamed constantly about their camp and stared in an insolent manner at the escort. Still more amazing was the large number of Buangtheuva's men who had come in during the evening and camped outside the southern extremity of the village. Worse still, the *mantri* of Buangtheuva refused to meet McCabe.[245] All this revealed that a spirit of defiance had spread throughout the eastern Lushais. These movements at any rate led Tytler to apprehend an attack and he accordingly began to take precautionary measures by posting double sentries around the camp.[246] The Lushai system of night and pre-dawn attacks were well known. But nothing happened that night.

At about seven that morning, Lalburha's *mantris* reported again that the chief would like to see McCabe provided he come out of the village without an escort. But McCabe insisted that the chief should come in and meet him.[247] If Lalburha failed to do so by nine he would collect all supplies of paddy, and occupy the village until his orders had been complied with.[248] Neither McCabe nor the Lushais waited till that time. Lieutenant Tytler was ordered to collect the paddy from the

---

244 *Ibid.*
245 *Ibid.*
246 *Ibid.*
247 *Ibid.*
248 *Ibid.*

village and the Lushais too began to appear armed with guns in the jungle adjoining to the village.[249]

Sesawng village was among the larger Lushai villages. It extended for about one thousand yards, running almost due north and south of the crest of a hill between the rivers Tuirial and Tuirini. Both the northern and southern ends of the village were about two hundred and fifty feet above the centre where the forces had encamped in the midst.[250] Since these two extremities were of strategic importance both the parties tried to gain control over it. Lieutenant Tytler immediately sent twenty men each to hold these important points.[251] The Lushais pushed in about three hundred warriors to pre-empt the British and immediately seized the northern crest. But the British drove them back to the jungle.[252] Apart from this attempt the Lushais managed to creep up close to the camp on all sides, and taking advantage of the cover provided by the houses and ditches, began a steady fire on the main camp. In the meantime, the signallers under heavy fire helio'd to Captain Loch of the Aizawl camp for reinforcements and ammunition.[253] The message was hardly finished when the Lushais set fire to the houses all around the camp.[254] After a brief confusion the men and the coolies collected themselves. The baggage and ammunitions from the houses were brought out and collected in the open ground in front of the chief's house.[255] After the flames had subsided a little, the men joined Surgeon Captain Melville on the northern crest, a place which was considered safer and where they could hold out against any number of Lushais as long as their ammunitions lasted.[256]

---

[249] Reid, *op.cit.*, p. 30.

[250] FEAP. December 1892; Nos. 42-46.

[251] *Ibid.*

[252] *Ibid.*

[253] *Ibid.*

[254] *Ibid.*

[255] Reid, *op.cit.*, p. 22.

[256] FEAP. December 1892, Nos. 42-46. McCabe to Secretary to the Chief Commissioner of Assam, 23 July 1892.

In the first attack a Jemadar and a coolie were wounded while the Lushais lost several killed and wounded and five guns. The firing of the village compelled Lalburha's men to confine their attack in the north and north-west of the British position. This enabled McCabe in the afternoon to regroup his men, make arrangements for water supply and bring in the six coolies and fifteen military police whom he had sent to Tuirial for more supplies. His position was strengthened by the arrival at 2:45 a.m. in the following morning of a hundred rifles, ammunition and supplies from Aizawl.

Meanwhile the Lushai warriors had not remained idle. Their forces too had been reinforced by four hundred men from Pawibawia's village. The attack thereafter remained intense till darkness set in. Throughout the night a desultory fire was kept up but they failed to prevent the Aizawl reinforcement from reaching McCabe. This was largely due to the fact they had busied themselves in constructing two stockades close to the village. By dawn these were ready; from there they recommended their firing. But with the reinforcement McCabe was able to dislodge them. It was now clear that resistance was not confined to Lalburha. As McCabe saw it:

> The presence of Poiboi's, Bungteya's, and Lenkham's men pointed to the fact that this attack was not a mere emeute on the part of Lalburha, but was undoubtedly a concerted rising of the Eastern Lushais. We were thus enabled to gauge the strength of the combination against us.[257]

As he could place no reliance on the western Lushais he asked for 300 military to be sent to Aizawl which would at the same time set free the 400 Military Police for operations against the Eastern Lushais.

Military operations at a time when the rains were approaching caused considerable worry to McCabe. Regular

---

[257] FEAP. December 1892; Nos. 42-46.

supplies from Aizawl could not be expected owing to the fact that after the rain set the paths would become impracticable. Added to this was the difficulty of transport. It was necessary to military operations independence of supply and transport from Aizawl. After consulting Captain Loch and Lieutenant Tytler McCabe drew up the following plans:

1. To fortify the position at Lalburha, so that it might be held by a small garrison.
2. To collect and store at Lalburha three months' supply of paddy for the sepoys and coolies, and so minimise the carriage of rice from Aizawl.
3. To make a good bridle path from Aijal to Sonai and establish a post there.
4. To try and induce Pawibawia's men to sever their alliance with Lalburha, and thus diminish the chance of raids in the plain.
5. To send out parties daily to hunt the enemy down from one hiding place to another, and bring home to each individual Lushai a knowledge of the hardship that such a rising entails.[258]

Captain Loch was directed to inform the western Lushais that none of their men were to cross the Tuirial or allow eastern Lushais to seek shelter in their villages. They were also to report all information they might obtain regarding the whereabouts of those whom the British considered their enemy.

McCabe's position in Sesawng village was quickly fortified: a stone wall was built on three sides, and the fourth was protected by a palisading. A godown was constructed, paddy collected and stored, temporary barracks and hospital erected and latrines dug. The water supply was thoroughly cleansed

[258] *Ibid*. This therefore minimised the work of *coolies* of the Public Works Department.

and Surgeon Captain Melville took every measure for the sanitation of the camp. During March 1892 the bridle path between Aizawl and Tuirial was taken in hand and the camp at the latter place garrisoned by 50 Military Police. All that remained was the arrival of the military to relieve the Military Police at Aizawl and Changsil.

LUSHAI RESISTANCE INTENSIFIED

Tytler posted daily pickets on the ridges towards Buangtheuva and Pawibawia, and sent out parties to hunt down the Lushais and collect paddy and other supplies. The pickets were often attacked with casualty on both sides. Sorties from the camp managed to disperse the Lushais who carried off the killed and wounded.[259] Two days later the Pawis of Liankhama's village made a determined attempt upon the British position. They occupied the southern point of the village commanding the line of communications with the Tuirial post.[260] After an intensive exchange of fire two Lushais were killed, several wounded and one sepoy mortally wounded. It was with great difficulty that the Lushais were dislodged. The following day, March, a counter-ambuscade under Subedar Jangbir surprised the Lushais at dawn, and forced them to retreat with a loss of one man and another wounded.

On the same day, the Lushais attacked a convoy of 62 coolies carrying ration under an escort of the Military Police, on its way from Aizawl to Sesawng Fort.[261] The convoy managed to reach the Fort but not before two sepoys were killed. The pickets around the Fort and convoys were subjected to constant attacks. Small parties of Military Police too constantly sallied out to destroy all Lushai temporary camp and stockades and collect or burn all the paddy in the *jhum* houses within a radius of five miles of Sesawng. From friendly or captured Lushais details of Lalburha's plans soon became

[259] *Ibid.*
[260] *Ibid.*
[261] *Ibid.*

clear. His warriors wished to build stockades on the road between Tuirial and their village. But the chief was reported to have said: "No! better let them settle down quietly in camp, invite the sahebs without an escort to a meeting outside the village, cut them down, and then in the confusion attack the sepoys and coolies". These plans fell through. But the lessons of the September 1890 rising, when repeated attacks upon Aizawl and Changsil posts had proved a failure, were never quite forgotten. McCabe thus describes Lushai tactics:

> No attempts were made to attack any of our stockades, but by spreading themselves in small parties along our line of communications and carrying on a guerrilla warfare the enemy succeeded in inflicting considerable loss on us. They hoped by killing a number of our coolies so to demoralised the rest that, being deprived of our transport, we should be compelled to retire.[262]

These tactics having also failed a large contingent from the villages of Lalburha, Pawibawia and Maite descended upon the plains and devastated the Boruncherra tea estate in the Hailakandi subdivision of Cachar on 4 April 1892.[263] Forty

[262] *Ibid*. Between 18 March and 9 April, British loses were summarised as follows:

18 March	-	a bullock driver was killed within 300 yards of Aijal,
20 March	-	a bullock driver was killed near Changsil.
23 March	-	a Military Police Havildar was wounded at Khama (post on the new Aijal and Changsil road.
27 March	-	A *sepoy* of the 18th Bengal Infantry was killed on the Aijal – Sonai road, and a police sepoy on escort duty with a bullock convoy was killed at Thanruma.
29 March	-	a police sepoy was severely wounded on the Aijal – Sonai road.
1 April	-	One police sepoy was killed and three coolies wounded on the Aijal – Sonai road.
2 April	-	A police sepoy was killed on the Aijal – Changsil road.
3 April	-	A sepoy of the 18th Bengal Infantry was severely wounded on the Aijal – Changsil road, and a police slightly wounded on the Aijal- Sonai road.
5 April	-	A police sepoy was killed on the Aijal – Sonai road. In this attack the enemy lost one killed and five wounded.
7 April	-	a police sepoy was slightly wounded on the Lalburha - Sonai road.

[263] Reid, *op.cit.*, p. 31.

five persons were killed and thirteen carried away. McCabe was convinced that the object of the raid "was to obtain heads and induce us to withdraw our outpost from Lalbura". The incident introduced a new dimension to Lushai resistance.

Raids on the plains had now to be considered even though it seemed improbable that during the rains the Lushais would dare to raid any frontier village or tea garden. But the natural condition of the country were all in their favour. The route of a raiding party lay through an uninhabited tract of country. All they needed to do was to march along the Rengti road to Kolosib, and then branch off along a jungle track to the defenceless tea gardens and border villages without incurring the risk of having their movements reported.[264] Again marching from the village of Sailianpuia west of Dhaleswari river to the border of Hill Tippera, a journey of ten days, not a single Lushai village could be seen. And if the facility with which the Lushais moved through the jungle independent of paths, it seemed surprising to McCabe that the frontier tea gardens had so long enjoyed immunity from raids.[265] The frontier of Sylhet and Cachar towards the Lushai hills was so extended that multiplication of the frontier outposts could hardly make raids impossible.

McCabe therefore firmly believed that instead of increasing the number of out posts along the frontier, efforts should be directed towards obtaining a thorough control over the Lushais.[266] "The history of the Sibsagar frontier," he reminded his superior in Shillong, "clearly demonstrated that the only successful method of preventing Naga incursions was to obtain a thorough control on these tribe by establishing posts on the heart of the country."

---

[264] FEAP. December 1892; Nos.42-46.

[265] *Ibid.*

[266] *Ibid.*

OPERATIONS

Meanwhile on 19 March the last detachment of the 18th Bengal Infantry arrived at Aizawl. These and the Military Police were disposed as follows:

Military Police			
Lalburha Fort at Sesawng	-	100	200
Tuirial Camp	-	50	
Sairang and Thanruma	-	50	
18th Bengal Infantry			
Aizawl	-	200	300[267]
Changsil	-	100	

This distribution placed 325 Military Police and seventy five Military at McCabe's disposal for the operations. By 3 April the Aizawl-Tuirial bridle path was completed in spite of almost daily attacks by the Lushais. Between 4th and 9th April fifteen days' supplies for 400 sepoys and 250 coolies was collected at Lalburha Fort and another 15 days supplies at the Tuirial Camp.

But before proceeding against Lalburha McCabe had to ensure that the western chiefs gave no trouble in his absence in the east. Though they strongly protested their loyalty to the British Government they told McCabe that Lalburha had threatened to make war on them if they did not join in the uprising and that they feared the result of their refusal. The barrack site of Lalburha's village and the Union Jack flying over the Fort was pointed out to them. The hint was clear enough. Nonetheless McCabe had no doubt that the "western Lushais desired to stand aloof and watch the result of the rising," and he considered it necessary to make them clearly understand "that we could only recognise open foes, or allies who proved their loyalty by implicit obedience and substantial assistance." Assistance came in the form of supply of labour and in clearing the jungle on the Rengte road.

After leaving Colonel Rennick in command at Aizawl McCabe and his force advanced towards Pawibawia on 10 April. As he approached the crest of the village the Lushais opened fire from two strongly constructed stockades.[268] After a short but intense exchange of fire, the main stockades, which completely barred the entrance to the village was stormed by the Military Police. The Lushais dispersed for cover in all direction, but were able to set fire to the connected village of Lalruaia burning all the houses. The main village of Pawibawia was also fired but the force managed to save a number of houses sufficient to provide shelter for the whole force.

McCabe remained in the village sending small parties to hunt down and destroy the camps in the jungle. A systematic search was made for concealed grain and about 3500 maunds were discovered. The whole country for a radius of ten miles from Pawibawia village* was thoroughly searched, camps were destroyed, and a large quantity of livestock and household effects were brought in.[269] This forced a large number of Lushais to seek shelter in the distant villages.

The result of this ruthless policy ensured the submission on 20 April of Lalhleia and forty-two of his warriors. This was followed by Selvunga's surrender the next day. But Pawibawia's widow with her two sons Lalruaia and Thangkama, fled to Saitual east of the Tuirini river. McCabe detained the surrendered chiefs as hostages but released all the others and settled in their *jhums* on the Hmuizawl hill.[270] He next advanced to Buangtheuva on 6 May. The Lushais did

---

[268] *Ibid.*

*Lushai villages are generally known by the name of their chiefs, hence Pawibawia or Pawibawia village, even though that chief was long dead. In the early maps of the Lushai hills the name of a village is usually shown by the name of the chief.

[269] *Ibid.*

[270] *Ibid.*

not contest his march for the first thirteen miles. Four strong stockades had been erected in the last three: the first two stockades were unoccupied but the third, which was built on the crest of the ridge commanding the village, was defended. They were soon overpowered and two Lushais were killed and one gun captured. An attempt of the Lushais to set fire upon their own village failed and McCabe obtained "excellent quarters up to our final departure on the 1 June".[271]

From Buangtheuva small parties continued to hunt down the Lushais in the countryside. On 9 May Lalthangvunga and the headmen of Maite surrendered.[272] Captain Loch with 100 Military Police, occupied the village of Chinhleia, Lalburha's half brother, on the 12 May. Two days later, Pawibawia's two sons Lalruaia and Thangkama, surrendered themselves. On 19 May Vanphunga, Pawibawia's eldest son came in into the Buangtheuva camp.[273]

Lalburha alone among the chiefs remained defiant, firm in what McCabe called a "no surrender policy." Along with a small contingent from Liankhama he resolved to make a stand at Maite, his mother's village. It soon became the centre of large number of refugee warriors from Buanghteuva, Lalburha, and Chinhleia, all of them acting on Lalburha's orders. Lalburha's objective was to prevent the advance of the British force beyond Buangtheuva's village.[274]

A force of 170 sepoys under the command of Captain Loch accompanied by Tytler, Edwards, Johnson, Surgeon Captain

---

[271] *Ibid.*

[272] *Ibid.* From then McCabe learnt that, "Lalburha subsequently organised raiding parties to harry the line of communication between Changsil and the various out posts, and the village of Maite supplied a contingents of 200 men with 40 gun for this purpose. Other Maite men were employed in carrying rice for Lalburha's people, and a party of 25 men with guns under a chief named Lalthanhranga joined in the raid on Boruncherra."

[273] *Ibid.*

[274] FEAP. August 1892; Nos. 203-214. Telegram from Political Officer, Fort Aijal to the Secretary, Assam, Shillong. Dated 29 May 1892.

Melville and Evans started for Maite. The force took the detour route through Saitual for they received the information that Lalburha had made elaborate preparations to make a stand on the direct road.[275]

The advance of the force on Maite from the unusual route took the Lushais by surprise and they had time to erect only one stockade on the ridge of the Saichal hill. The village was captured without difficulty. On the evening of 25 May the Lushais began a determined attack and maintained it till dark. In the days that followed search parties were again organised to comb the countryside. On the 28th Lieutenants Edwards and Johnson surprised a camp of the Lushais at old Maite village, while other parties destroyed the stockades erected on the direct road to Buangtheuva.[276] The operations were now at an end with Lalburha and about twenty faithful followers, fleeing to the impenetrable jungles of the Manipur frontier.[277] The headman of Liankhama submitted to the Political Officer of the Burma column.

After initiating the process of resettlement of the scattered eastern villages the punitive force returned to Buangtheuva on the 30th, and to Lalburha Fort on the 1 June. A week later McCabe and the whole force returned to Aizawl.[278] In his report on the operations McCabe observed:

> I trust the lesson which has been taught is one not likely to be forgotten. The opportunity of a stand up fight and the infliction of a decisive blow is never offered, and in the petty guerrilla warfare, which was to be carried on, the advantage are all on the side of the enemy. It is only by the destruction of their food-supplies and by preventing them from jhumming that we can hope to

---

275 This would also enable the British to ascertain the feelings of the Pawibawia group of villages. The friendly reception Loch received at Saitual clearly indicated that Lalburha was now totally isolated.

276 *Ibid*.

277 S K Barpujari, *op.cit.*, p. 242.

278 FEAP. December 1892; Nos. 42-46.

> enforce an unconditional surrender–means which, though appearing somewhat harsh, are in the end the most humane.[279]

Neither did other officers forget the efficacy of McCabe's methods, as events in the South Lushai Hills, even as Captain John Shakespear was developing his own system of control, were to show.

---

[279] *Ibid*. The Chief Commissioner agreed: "So far as it is possible to predict anything of these hill tribes there seems to the Chief Commissioner to be every ground for believing that the Lushais east of the Sonai river have been as completely crushed as the western Lushais were by the expedition of 1890-91, and they are not likely to give much trouble in the future if proper outposts are established and maintained in their country." F C Daukes to Secretary, Foreign Department, 24 October 1892.

4
PROBLEM IN SOUTH LUSHAI HILLS

After its attachment to Bengal South Lushai hills was under administrative control of the Deputy Commissioner Chittagong Hill Tracts. C S Murray of the Bengal Police and Captain John Shakespear of the First Leinster Regiment, both of whom had distinguished themselves in the Chin-Lushai Expedition, looked after its affairs as Assistant Political Officers.* While McCabe was involved in crushing resistance in the north, Shakespear and Murray were engaged in the pacification of the south. The impetus to the annexation of these hills had come from raids that the tribes in the Lushai hills made on British or protected villages. But within the hills the situation was extremely complicated. Inter village feuds had been rampant and pacification involved putting an end to this and enforcing *Pax-Brittanica.*

During January and February 1891 Captain Shakespear from his base at Fort Tregear moved eastward visiting village after village imposing fines and releasing captives. Every village was made to provide supplies and coolies for transport of baggage. By early February he reached Saiha after a strenuous march. Here he found that the "language of these

*In November 1890 Bengal decided upon a Superintendent and had actually appointed Murray to the billet. It was however, not before April 1891 that the South Lushai Hills was constituted separate district that a Superintendent could take charge.

people is quite different to that of the Lushais or eastern Pois." Saiha was in fact a Mara village, a tribe known to the Lushai and the British as Lakher and to the Arakanis and others as Shendus. So too was Ramri village which he next visited. The state of affairs in the Lakher country in the south east of these hills can be seen from the nature of Ramri defences described by Shakespear:

> The defences have been very elaborate. The approach from this side is along a narrow ridge. At the end of this ridge the naturally steep hills has been scraped away and at the top is the gate way, the only approach to which is up a single log knotched to form a rough ladder. There have been two "keeps," one having a stone wall four feet thick surmounted by a palisade, but both are in a very bad state of repair. The chief came to greet us, and conducted us to two nice houses at the top of the village.[280]

By the middle of February Shakespear reached Serkawr one of the last of the Lakher villages towards Burma. Compensation was assessed and imposed on Serkawr, as was done in the case of Saiha, for raids which were guilty of. From Ramri captives taken from northern Arakan villages were recovered. It was while in these parts that Shakespear learned of Murray's difficulties in Zakapa's village.

SUBJUGATION OF THE PAWIS

C S Murray, along with S V Tayler, had moved out of Lunglei on 1 February 1891 with fifty Chittagong Frontier Police on an extensive tour.[281] On the 8th he reached Zakapa's village where he halted the night.[282] Zakapa was apparently friendly and promised Murray cooly transport for him to proceed to Lalbuta's village some miles to the north. The following day Zakapa was sent for along with Lalchhuma, a

---

[280] FEBP. May 1891; Nos. 56-61, Diary of Shakespear, 5 February 1891.

[281] FEAP. July 1896; Nos. 71-115, Diaries of C S Murray, from 1 to 18 February 1891.

[282] *Ibid.*, C S Murray to Commissioner of Chittagong Division, 12 February 1891.

neighbouring chief whose village was only two hundred yards distant. Both of them, as Murray noted in his diary, "refused to come" on the ground that they were busy organising the coolies.[283] On the 10th they were again summoned but with the same result. Thereupon the Assistant Political Officer marched to Lalchhuma's village to find them in the chief's house with a large number of men around them. Murray ordered the men out and asked the chiefs to tell him once for all whether they would obey his orders or not.[284] He threatened to punish them severely unless they followed him to Zakapa's village. But once outside the house they were prevailed upon by their men not to proceed. Before leaving Lalchhuma's village Murray turned and told the chiefs: "this is the last time I give you the order; if you refuse to obey, you must take the consequences."[285]

From Zakapa's village Murray leaving twenty men behind moved out with the remainder of his escort to destroy thirteen grain silos some two and half miles from the village. While these were being fired the Lushais turned on the small guard in the village and looted the government chest of nearly Rs. 2000. After this they ambushed Murray on his return:

> We hurried on as hard as we could, (Murray wrote) and when about three-fourth of a mile from the village, where the path was narrow with thick jungle on both sides, a volley was fired into us. I had told off an advance guard of five sepoys and one havildar, who were about five paces ahead of me. The two leading men were shot dead; the third man had his Frontier Police badge shot off his cap; the fourth man was untouched; the fifth man, the havildar, had his cap shot off; the sixth man in front of me had a bullet pass through his belt and clothes behind, just grazing his back, two bullets hit the muzzle of his rifle about three inches and twelve inches from the muzzle of his rifle being on the slope.[286]

---

283 *Ibid.*

284 *Ibid.*, Diaries of C S Murray, From 1-8 February 1891.

285 FEAP. July 1896; Nos. 71-115, Diaries of C S Murray, from 1 to 18 February 1891.

286 *Ibid.*

Both Murray and his assistant Tayler were singled out for attack. As Murray continued:

> A second later several guns were pushed out of the jungle, and deliberate shots taken at Tayler and myself. I do not know how we escaped, as the guns were not three feet off us. A volley was at once fired by our side, and I gave the order for all of men to lie down in the jungle which we did. I had heard shots continually from the village up to the attack on us, when they had ceased. I ordered the bugler now to sound the "fall in," to try and let the police in the village know where we were. The result of this was to bring a volley in the direction of the bugler. I now held a consultation with Tayler and my native Officers, and we came to the conclusion it would be madness to try and advance. The Lushais were evidently in large numbers. They had plenty of guns, absolutely commanded the position in every way, and we could not see one of them:[287]

Murray and his men managed to retreat and made their way to Chhimtuipui river and reached the camp on the following day at about four in the evening.

The incident brought a sharp reaction from David Robert Lyall, who as Commissioner of the Chittagong Division, was Murray's immediate superior. He at once asked Murray to explain the entire episode, particularly on the alleged refusal of the chiefs to provide coolies. It appeared strange to him that Zakapa, who had earlier sworn friendship with Captain Shakespear and was apparently friendly towards Murray at first should suddenly and for no apparent reason attack him in strength. That Lyall's suspicions were not without foundation was soon established when it transpired that it was not the demand for coolies or supplies that was at the root of the attack upon Murray. In the enquiry conducted subsequently Murray's interpreter deposed as follows:

---

[287] *Ibid.*

> Jacopa sent 50 coolies to carry the baggage. Mr.Murray had asked for 100. On the arrival of Mr. Murray to Jacopa's village, Jacopa met the Saheb with welcome. Mr. Murray told me to get two girls for himself and the Chota Saheb (Mr.Tayler). I said, where shall I get them from. Mr. Murray said, tell the chief to get them. I did so. Then I and Vaitlaia and Jacopa searched for these girls, and could not get them any. Then we went to Mr. Murray and told him and he said 'Why cannot you get them; go and make a 'Banda bust'. Afterward we persuaded two girls to come but when I told them that they would sleep with the Saheb, they said "No, we won't" and run away.[288]

The interpreter returned to report their desperate search for girls and asked Murray that they will catch them if he agreed. This was not agreed to as Murray clarified his position:

> No do not catch them, the Sepoy will see, but if you cannot bring me two women, I will have the wives of Jacopa and Pajika. Jacopa was present and Mr. Murray said it in Lushai, and at once Jacopa's family began to leave the village.[289]

In all probability all this happened on the first night itself. This explains why Zakapa and his family moved out to Lalchhuma's village that very night. Murray himself noted in his diary that Zakapa "did not sleep in the village this night." The demand for girls was again made to Zakapa in Lalchhuma's village according to the interpreter. Further he asked the chief to collect the village girls to dance before him to which Zakapa replied that according to their custom this could be done only on traditional festive occasions. Zakapa instead offered him goats as a gift, but when these were supplied Murray had found fault with them. Incidentally, Murray in his diary refers to the refusal of the chiefs to his demands, but cleverly does not state what these demands were. The Government of Bengal accepted the enquiry report

---

[288] *Ibid*. Statement of Panjiham Tipperah, son of Chandra Singh, an Interpreter.

[289] *Ibid*.

and recognised that Murray's "disgraceful conduct was the principal cause of the outbreak." He was immediately removed.

News of the outbreak and particularly the reason behind it soon spread to other neighbouring villages.[290] Fifty nine men from Dokapa's village who acted as Murray's carriers up to Zakapa's village and were on the way to Lalbuta's village threw down their loads when they learnt of the incident and joined in the attack.[291] From Dokapa's village as well as from Kapchhunga's and Lalchhuma's quickly joined in.

On the 22 February Captain Hutchinson with 150 men advanced from Fort Tregear to Zakapa's village.[292] By the time he reached the village, Zakapa had already moved out and remained moving from place to place to avoid an encounter with the force. All military and police post in the hills up to Fort White was telegraphed to obtain intelligence on Zakapa's movements.[293]

Meanwhile flying columns under Colonel Evans, Lieutenant Cole, and Captain Hutchinson scoured the hills hunting the chiefs and their *mantris* involved in the late rising. By early March three chiefs, Dokapa, Lalchhuma, Kapchhunga, and their *mantris* were apprehended. Zakapa still eluded the British forces and remained a fugitive. It was not before another four years that Zakapa could finally be brought in.[294] After Captain Shakespear took charge as Superintendent on 16 April,[295] he announced rewards of

---

[290] *Ibid.*

[291] FEAP. July 1896; Nos. 71-115, (No. 83) Dairies of C S Murray Esq. C.I.E. Assistant Political Officer, Lushai Hills, From the 1st to 18th February 1891.

[292] *Ibid.*, No. 72. Confidential from Sir John Edgar to The Commissioner of Chittagong Division. McCabe also moved towards the South from the North Lushai Hills.

[293] *Ibid.*, Report from C S Murray to the Commissioner of Chittagong Division. 8th March 1891.

[294] *Ibid*

[295] *Ibid.*

Rs.500/-, Rs.200/-, Rs.100/-, for the capture of Zakapa, his brother Pazika, and the *mantri* respectively.[296]

The subjugation of Dokhama, Lalchhuma, and Kapchhunga brought peace to the southern hills for the remaining months of the year. On 4 January 1892, Shakespear held a *darbar* of chiefs near Lunglei which was attended by representatives from every tribe.[297] The Superintendent told them that British occupation was permanent and that feuds unless stopped would involve serious punishment. They were made to swear friendship with one another or at least peace with one another. Tribute would have to be paid, and Shakespear accepted the principle that this could be done in the form of rice. He also told them labour supplies would be paid for, the amount later being fixed at four annas a day.

At the end of the month Shakespear met McCabe at Kairuma's village and settled the boundary between the North and the South Lushai hills. From here he proceeded south again to enquire into a murder committed by the Pawi chief Dokulha, a brother of Hausata involved in a killing of Lieutenant Stewart that had occasioned the Lushai Expedition of 1889. Dokulha's statement which he made two days after his capture illustrated the nature of the feud in these hills:

> Thongliena's men shot my brother Vantura. If I did not kill some men my brother's spirit would have no slave in the "Head men's village, therefore I went and shot two men of Thongliena's village. We met some men of Boite Thilkara's village and mistook them for Thongliena's men and so shot at them.[298]

On 17 February Shakespear sent his assistant Robert

[296] FEBP. August 1891; Nos. 88-97, (No. 93) Dairy of Captain Shakespear from 23rd to 29th May 1891.

[297] Sir Robert Reid, *History of the Frontier Areas Bordering on Assam*, (Shillong, 1942). p. 45. Five clans represented this *darbar*, "*Howlong, Thangloa, Mollienpui, Lakher* or *Longsen*, and *Poi*."

[298] Quoted in *Ibid.*, There is a misprint; it should have been "Dead men's village" instead of Head men's village. "Dead men's village or *Mitthi Khua*" is the place where the soul of the dead went.

Sneyd Hutchinson with thirty-six men under a Subedar against Dokulha's village. Hutchinson started at 3 a.m., and in about an hour's time reached the *jhum* fields of the village. While near the village on top of the summit of a hill the alarm was given. Hutchinson thus described what followed:

> I charged into the village with some fifteen men who were near me; men with guns came tumbling out of the houses and I heard shots fired. I had ordered my men not to fire but to follow me in a rush on the chief's house, the situation of which I knew. Unfortunately a man with a gun took deliberate point blank aim at me and I fired at him with my pistol, he lurched forward dropping the gun but was seized and carried off by some other men near him, the gun remaining with me. The delay of a minute or so just stopped me from getting Dokola who made away as I entered the house in company with some other men.[299]

The man who disappeared before Hutchinson was none other than Dokulha himself. The next day, 18 February, Dokulha gave himself up. This was just as well for disturbances had now broken out to the north.

THE HAULAWNGS

In February 1892, it will be recalled, began Lalburha's resistance against McCabe. Shakespear received a message from Aizawl that McCabe was closely invested by large number of Lushai warriors at Lalburha's stockade. Shakespear at once decided to proceed to McCabe's assistance. The latter however, when asked reported that assistance from Lunglei was not required but that if Shakespear could move around the Haulawng group of villages it would prevent them from joining the eastern Lushais. The villages of these chiefs lay between the north and south Lushai hills. The Haulawng chiefs as Shakespear knew were a branch of the Sailos and therefore

[299] Quoted in *Ibid.*, p.33.

connected with Lalburha and the eastern chiefs.[300] The most important of them were Dokhama, Dophunga, Lalluauva, Seipuia, Vansanga, Lalhrima, Lalthawma, Rochungnunga, Tlungbuta and Lalburha.[*] Lalthuama whose village was situated nearer Lunglei was in fact under the influence of Ropuiliani. She was the wife of Vandula and daughter of the great chief Lalsavunga. Of her Shakespear had written:

> Her influence is directly hostile to us, as is only natural when it is considered that she is the daughter of one great chief who always opposed us, and the widow of another. Since her husband's death she has seen his brother becoming more and more friendly with us, and increasing their prestige by virtue of this alliance. So much so that I overlooked her entirely and attributed Lalthuama's frequent faults to his own youthful folly. All the villages belonging to this group have been more or less troublesome, not actively hostile but passively obstructive. It has always been difficult to get tribute or labour from them.[301]

Ropuiliani had tremendous influence among the Lushai chiefs, who were by connection sympathetic with the eastern Lushais. Lalluauva and Seipuia were however friendly towards the British.

---

[300] Of them Shakespear had thus recorded in his diary for the week ending 8 August 1891, the entry being for 6 August: "Apparently the people known to us as Howlong are really a branch of the Syllo clan and are called Howlong after the hill of that name on which the village of Rollura was the same stock as the chief shown in the pedigree tables as Sailos. The whole of the tribes known to us as Sylos and Howlong are spoken of the common term Thang-ur."

[*] Others were: Kamlova, Lalkanglova, Zaduna, Kaphleia, Kalkhama, Lalmaia, Laltawna, Liankhara, Lianpuiluti (mother of Laltawna), Sailoliana, Vanthanga and Thangliana. Lalburha who was involved in this rising was not the same man with that of the Eastern chief. He was, however, usually referred to as South Lalburha.

[301] FEBP. March 1894; Nos.152-159, Shakespear's report concerning Ropui Lieni, widow of Vandula, and her son, Lalthuama.

[*] Intelligence of Lushai plan and movements were given by Dara Shakespear's friend and interpreter. For a photograph of his, in old age, see, A.G.McCall, *Lushai Chrysalis* (London , 1949), facing p. 64. The success of the forces however, rested on Dara, who explained the Lushai mode of warfare and their strategy.

Shakespear's objective in advancing into the Haulawng region was to ensure the neutrality of these villages. He was also to prevent Kairuma from making common cause with Lalburha. On 13 March Shakespear reached the friendly village of Lalluauva. Here he discovered that all the villages decided against supplying coolies. He was further told that the Lushais were prepared to resist his advance. The first indication of this was the killing of Shakespear's interpreter Satinkhara, at the village of Dokhama, the grand son of Ropuiliani. On 16 March Shakespear was himself attacked near Vansanga's village of Chhipphir. This was repulsed. But Shakespear now realised his difficulties. The terrain and the dense grass jungle along this route made it difficult even to see the attacking Lushais. An attempt at negotiations failed and Shakespear had to push on. Every inch on his passage was contested and it was after great difficulty that Vansanga's village was reached and carried. It was now fortified but remained virtually under siege by the Lushai.

Vansanga's mother who ruled Zote village called Shakespear to her village so that talks could begin with her son. But on his way there Shakespear was attacked, and had to return. After collecting supplies he moved out against Zote village on 18 March with hundred rifles and 190 coolies, a guard of fifty rifles being left to protect his base at Vansanga. As he neared Zote village Shakespear found it was strongly defended.[302] Not less than four stockades blocked the path to the village and were defended by fifty to sixty armed men. It was clear that the Lushais were determined to block the movement of the force beyond Zote.

> I found Zaute strongly stockaded and full of men, I stormed the stockades, losing one sepoy. The rebels holding the stockades till we were within a few yards of them. It was now evident that 150 rifles could not quell the rebellion. I therefore returned to Vansanga's and stockaded my force...[303]

[302] MSR. General-9, CB-1.

[303] *Ibid.*

The incident at Zote was the result of a confederation comprising men from Thangliana, Kamlova and Lalhrima and some other village.[304] Shakespear now realised that he could not occupy Zote with a small force. There was constant sniping from the jungles and he was short of supplies, not having even the unhusked grain. The village was accordingly fired after which Shakespear returned to Vansanga.[305] From here he sent small flying columns by night to nearby villages.[306]

Shakespear could neither move forward nor control the developing resistance. The Lushais cut the telegraph line, as a result of which no communication to any other place was possible.[307] Meanwhile simultaneous attacks were made by the Lushais on the Vansanga position. Shakespear had no choice but to seek reinforcement from Chittagong. He sent Lalluauva's men thither with the following message:

> I am sure from the determined resistance made today that the rising is a big one. At present our position is most precarious, and any hesitation now would be fatal. It is only by holding on here and putting a bold face on the matter that we can succeed; but full and complete punishment must at once be meted out, and a complete submission obtained. I cannot do this without more men.[308]

---

304 *Ibid.*

305 *Ibid.*

306 As this village stood on the top of the cliff it was convenient for the Lushais to take defensive position and they stockaded the village, properly placed the fully armed sentries.[3] Lieutenant Boileau and his party made a flanking attack, fired a volley and then charged up the hill towards the small gate, which was fortunately opened. At the same time the main body tried to divert the attention of the Lushais which enabled the flanking party to get into the gate, one *sepoy* killed and one Lushai from Lalluauva's men wounded. The Lungrang sentries were taken by surprise from the rear and they abandoned their position and frittered away to all directions. The village was then destroyed. On their way back to Vansanga's stockade the Lushais continued to fire from the jungle, in spite of which they managed to reach the village stockades, where the force was better protected from the Lushai continuous attack.

307 MSR. CB-1, Pol-3, Shakespear Note on the Lushai Hills, its inhabitants and its administration since 1888.

308 Quoted from John Whitehead, *Op.cit.*, p.108.

On 21 March, Hutchinson arrived at Vansanga's village with 58 rifles from Lunglei. Though this strengthened his position somewhat, Shakespear believed that it was still not possible to attempt punishing the hostile chiefs whose villages, about 20 in number, were scattered over several square miles of the territory.[309] Shakespear accordingly returned to Lunglei leaving a guard in Vansanga's stockade under a European Officer.[310] News immediately spread that he had retired and "immediately the Lushais swarmed round the post and the lines of communication, cutting the line, firing on all parties and cutting up stragglers or the unwary who strayed even a few yards outside the posts."[311]

It was now that the Government of Bengal informed Burma of the serious state of affairs in the Lushai hills and applied for assistance. At this time a column called the Nwengal Column under Bertram S Cary was conducting operations against the Chins immediately east of the Lushai hills. When he received direction from his Government that he should proceed to assist the beleaguered Shakespear, Cary was on his return to Fort Falam. His transport was worn out and rations all but exhausted, and the food supply taken from Nikuala village, his objective in his last expedition partially consumed. He therefore decided to return to Falam and fresh and replenished was back to the assistance of Shakespear.[312] After overcoming problems of moving in equally difficult terrain and under incessant rain the Burma column reached Dokhama village on 3 May. On the news of its arrival the Lushais ceased firing into the post, deserted their villages and began hiding their grain and property in the jungle. Cary discussed the situation with Captain Rose who commanded

---

[309] M SR, CB-1, G-9, J.Shakespear's report. 29th June 1892.

[310] *Ibid.*

[311] Bertram S.Cary and H.N.Tuck, *The Chin Hills*, Vol. I, p. 70

[312] *Ibid.*, "To right about turn and relieve Captain Shakespear" wrote Cary in a report on his Expedition, "was for the moment perfectly impossible. We were still four days' march from one outpost Botung, and had three days rations

the column and with Captain Shakespear. Since the column had rations for five days operations it was, "decided to punish a few villages, but to punish them thoroughly" and a campaign against the leading village was arranged. What the Burma or Nwengal Column did is described in Cary's own words:

> every nullah was searched and tons of grains burnt, whilst herds of cattle were shot and finally the village burnt...The Burma Column marched to Rochungnunga, shelled the village of Lalkanlova en route, and also destroyed granaries. The Frontier Police marched to Rochungnunga by a different route, and on arriving at the village were fired on by a few Lushais and a few shots were exchanged...The Lalrhema party slipped through it just before daylight, but the place was deserted except for a small guard with whom shots were exchanged for some minute: the party then rushed the village. The Lalkamlova party rushed the village as daylight was appearing, killed one man, wounded another, and captured a third, seized four guns, and, after burning the village to the ground, returned to Rochungnunga.[313]

So it went on all altogether 1500 houses were burnt. Thereafter the column moved on to Lunglei and thence to Chittagong and Rangoon.

RESULTS REVIEWED : OPERATIONS PROPOSED

After the subjugation of the Haulawngs and the punishment of a number of chiefs, Captain Shakespear reviewed the whole rising in detail. "The cause of the rising," Shakespear noted, "was the determination of certain powerful chiefs near Aizawl not to supply labour."[314] This he had experienced himself when in early January 1892, he sent out orders to the chiefs north of Lunglei and south of Aizawl, to

only. Eighteen miles behind us was Nikwe's village, which I had already fined all the livestock in the village and to have marched to relieve the Lushai force with three days' rations was out of the question."

313 *Ibid.*

314 FEAP. February 1893; Nos. 45-105, Shakespears' note on the rising among the Haulawngs. 5 September 1892.

send in coolies to work at the ordinary rate of pay. "They all made excuses," the Superintendent said and:

> I visited some of them, fined them for their disobedience, and by the middle of February had 300 men working at Lungleh. These men all ran away, being summoned by Vansanga and Lalthova. By comparing dates, it will be seen that just at that time, 19th February, Mr. McCabe was sending to demand coolies from Lalbura and was contemplating a refusal of that chief to comply with his demand.[315]

Shakespear firmly believed that this disobedience in recalling their men from their work at Lunglei was on the advice of the northern chiefs, who were subject to McCabe.

Captain Shakespear had further suspicion about this recall of labour. The nature of work was such that the hardship was slight. He knew that coolies were never called for except in the season of the year when there was no work to be done in the *jhums*. Besides, the men were paid for their labour at a fair rate. Furthermore, the work they were employed on was mostly house building, which the Lushais were accustomed to. On top of this the workers were found cheerful and willing. But some chiefs seemed to have been unhappy, Shakespear affirmed, "several chiefs asked last year why they alone were asked for coolies and not the more distant chiefs as well."[316] In spite of that, the Superintendent knew well, that the only way to reduce the cost of the British occupation of the hills was to obtain as much labour as possible locally.

Of the twenty one villages comprising some 3000 houses, concerned in the uprising, only nine villages had been punished; viz: the villages of Dokhama, Vansanga, Laltawna's mother's, Laltawna, Rochungnunga, Lalkanglova, Lalhrima, Tiungbuta and Sailoliana. Even these villages were not sufficiently dealt with as burning a village was but only a

[315] *Ibid.*
[316] *Ibid.*

"small punishment."[317] Shakespear therefore considered it absolutely necessary to send punitive expeditions against these chiefs. The mere sending in a military force was, of course, not sufficient to subjugate the Lushais. He explained what was needed:

> It has been over and over again conclusively proved that there is only one way to really punish these people, and that is to establish oneself in a village in sufficient force and with ample supplies, and regularly hunt them down until the terms offered were agreed to, and then repeat the operation at the next village.[318]

That burning of villages was not a sufficient deterrence was further elaborated upon:

> To burn the village is no punishment, as the best village in the hills can be rebuilt in three months. They will certainly bury their crops, so that a force having only time to march through the country will be able to do them little or no harm. In such a country there is no chance of bringing the enemy to bay and inflicting severe losses on them, and all Officers who have experience of the north-east frontier are agreed that the above is the only satisfactory method of obtaining a real submission.[319]

"I consider it absolutely necessary that a punitive column should be sent against the rebels this cold weather," he told J A Crowfort, Deputy Foreign Secretary, in a demi-official letter on 5 September 1892, and:

> It required all my influence to keep the remainder of my district quiet during the rising last spring, and if I am not placed in a position to coerce the rebels this cold weather, our prestige will suffer, so that most of the work of the last years will be thrown away and have to be re-done at an increased cost later on.[320]

---

[317] *Ibid.*

[318] *Ibid.*

[319] *Ibid.*

[320] *Ibid.*, See Note by Captain Shakespear, on the recent rising in the Lushai hills, and suggestions as to the operations necessary during the ensuing cold weather to complete the pacification of the District. 5 September 1892.

He considered that the whole operations would probably take about three months. Accordingly, he proposed to form a depot at Lalluauva's village, three days march from Lunglei, and collect two months' provision. The column would when near other posts be able to draw on the godowns there. He further planned to attack Vansanga's village and occupy it, collect as much rice as required and destroy the remainder, bring in all livestock they would find, and hunt down any of the rebels whom they might come across. After establishing itself at Vansanga's village, the expedition would proceed on to other villages and apply the same method. No offer of any terms or any overtures to the Lushais would be made unless it was in a position to take action should these be rejected. Shakespear had in mind what the terms and conditions would be:

1. The surrender of the chiefs, to stand trail.
2. The surrender of their guns.
3. Immediate compliance with all orders regarding supply of labour.[321]

If he was to achieve his objectives, Shakespear made the need for reinforcements clear. The punitive column would not be less than 350 rifles, a sufficient force would be required to guard the posts and to secure the Lunglei-Demagri line of communications. The total force required and their disposition was estimated as follows:

Fort Tregear	—	50 rifles.
Lalthuama	—	30 rifles.
Lunglei	—	150 rifles.
Demagri	—	100 rifles.
Six camp guards on Demagri – Lunglei road	—	180 rifles.[322]

[321] *Ibid.*
[322] *Ibid.*

The force available at his disposal was, 200 from 17th Bengal Infantry and 320 Military Police and Shakespear proposed another 400 men or a wing of regiment to be sent.

As the villages to be punished were so scattered even though small in number, the Superintendent thought the work would be simplified if a force from Aizawl could operate against the most northerly villages, *viz.*, Lalburha, Thangliana, Tlungbuta and Liankhara. The Lunglei column would simultaneously operate from the south. The other arrangements such as signalling, road-making, transport and supplies were also detailed. The date of commencement of operations would depend on the collection of sufficient supplies at Lunglei, but Shakespear hoped it would in the early part of December 1892, the best season for military operations in the North East Frontier hills.

Captain Shakespear had the full support of his Government. The Lieutenant Governor Sir Charles Elliott, however, considered it "absolutely necessary" for combined operations from Aizawl and Lunglei.[323] In Shillong Sir William Ward who discussed Bengal's proposals with McCabe agreed. The Political Officer of the North Lushai Hills reminded the Chief Commissioner that the Haulawng villages of south Lalburha, south Lalhrima, Tlungbuta, Thangliana, Kamlova, Liankhara, Kanglova, Rochungnunga and Zaduna which Shakespear had previously transferred to him all rose during the late uprising.[324] McCabe, further said that South Lalrhima, Kanglova, Tlongbut and Rochungnunga were destroyed by the Lungleh and Burma columns and were not likely to offer any organised resistance in the future. The establishment of an outpost at South Lalrhima, and a combined promenade through the northern Howlong country would undoubtedly effect a thorough settlement of this tract. He therefore suggested that:

---

[323] *Ibid.*, Demi official, Elliott to Lansdowne, 11 September 1892.
[324] *Ibid.*

1) The two columns should start simultaneously.
2) Vansanga should be the advanced base of the Lungleh column, and Leinpunga or Zahrawka of the Aizawl column.
3) The Aizawl force would proceed first to Lalbura, only two very easy marches from Leinpunga, thence to South Lalrhima.
4) The Lunglei force should visit Rochungnunga, Kanglova, Leinkhara and possibly Tlongbuta.[325]

By this plan, McCabe concluded, the country would be thoroughly patrolled, and all possibility of a combined movement on the part of the enemy would be frustrated.[326]

These recommendations were further elaborated by Bengal. Co-operation from Fort White or Falam was mooted, not so much to strengthen the punitive forces as much to block the escape of the recalcitrant Lushais from this area eastwards.[327] Columns from Fort White, Aizawl, and Lunglei should make a combined move from their respective position not later than 15 December 1892 towards the Haulawng country.[328] These three columns should establish heliographic communications with each other instead telegraph lines which were liable to disruption and so simplify the problem of information. They should arrange their own commissariat and supplies.[329] The column from the Fort White should cover the villages of Liankhama in the north through Nikuala's village, then Kairuma, Zaduna and Lalbuta in the south. The object of the operation would be to give protection to the friendly Lushai chiefs and consolidate the British power among the

---

325 FEAP. February 1893; Nos. 45-105, F C Daukes, Secretary to Chief Commissioner of Assam to the Chief Secretary, Government of Bengal. Dated 30 September 1893.

326 *Ibid*.

327 M SR, CB-2, G-25, H J S Cotton, Chief Secretary to the Government of Bengal, to the Secretary to the Government of India, Foreign Department. 6 October, 1892.

328 *Ibid*.

329 *Ibid*.

hostile. It was essential, in Bengal's view, that a sufficient force was established in the hostile villages and remained there until the terms were complied with.

Regarding the strength of each column, Bengal proposed that the Aizawl column should consist of 300 Military Police, and a column of not less than 300 from Fort White with two Mountain guns, and three to four hundred men of Military Police stationed at Lunglei and Fort Tregear would be sufficient.[330] In addition to these Gurkhas were asked for. They were considered experts in hill and jungle warfare and able to withstand the climate of the hills.[331] One other advantage of Gurkhas was that they were rice eaters like the Lushais and hence there would be no difficulty in finding supplies for them.[332]

The Government of India was reluctant to sanction the proposed extensive operations; Lansdowne's own view was that "considering the operations which we are carrying out elsewhere and the state of our finances, I should deprecate anything like unnecessary activity with Lushai Hills this winter."[333] Captain Shakespear who was in Calcutta tried to argue his case with the Military Member, Sir Henry Brakenbury. He explained to him the policy followed of compelling the Lushais to supply force labour and a certain quantity of supplies. Unless the policy was applied to the distant villages they would abandon their present homes and move further away. Brakenbury's views summed up the current thinking of the Government of India on the subject:

> I am not at present convinced that the policy of letting them all alone, and neither requiring force labour, nor tributes in supplies, might not be tried. I take it that these demands are the causes of the existing hostility to us on their part; and it might be possible to try the policy here of giving small subsidies for local levies and keeping

330 *Ibid.*
331 *Ibid.*
332 *Ibid.*
333 *Ibid.*, see KWs, Note by Lord Lansdowne, 25 September 1892.

> open lines of communication, which would probably be for cheaper on the long run than these continued expeditions.[334]

It was however on the insistences of Sir Charles Elliott[335] on punitive measures that the Government of India finally relented.[336]

COLD WEATHER OPERATIONS

In December 1892 combined movements of Military and Military Police of the strength already decided, began against the Haulawngs. A W Davis, who had in the meanwhile succeeded McCabe as Political Officer of the North Lushai Hills, with a force of Military and Militrary Police had already advanced in November as far as the village of south Lalburha, situated on the Darlung peak and containing 300 houses, and lying in his North Lushai Hills District. Lalburha surprisingly offered no opposition. He received Davis in a friendly manner, provided coolies for the transport of their baggage and whatever supplies he required. Still, Davis fined him 20 guns for his delay in providing the coolies.[337] This was promptly paid. Evidently Lalburha had no wish to fight the British though he was slow in obeying orders. From Lalburha's village Davis moved to Serchhip, Lalhrima's village. On the way he visited Samlukhai, the village of the Haulawng chief Thangliana, where he was well received. He reached Serchhip on 5 January 1893 and established an advanced post. Buildings for its permanent occupation by the Military Police, were constructed.[338] Besides with his base at Serchhip, the Assistant

---

334 *Ibid.*, Note by Sir Henry Brakenbury, 29 September 1892.

335 *Ibid.*, See Note by Lord Lansdowne, 26 October 1892. The Foreign Department also noted that the "Lieutenant Governor's meaning seems to be that, if the villages are left alone now, they are certain to give trouble which will force us to punish them within a short time..." and that, 'the credit of the British name' is doubtless at the button of the Lieutenant Governor's strong recommendation...", 6 November 1892.

336 *Ibid.*, Major General E H H Collen, Military Secretary, Government of India to Adjutant General.

337 MSR, CB-142, G-688, *Op.cit.*

338 *Ibid.*

Commandant of the Military Police, Lieutenant Tytler attained the submission of a number of chiefs.[339]

Captain Shakespear marched from Lunglei, commanding the combined forces from Chittagong and the force already at his disposal at Lunglei Fort. He covered the whole Haulawng villages towards Serchhip through Vansanga's village.[340] He encountered no opposition from the Haulawngs and made his way through to Serchhip village.[341]

Shakespear met Davis at Serchhip on the same day and the two Officers decided to impose a fine of 200 guns on the villages of Lalhrima, Rochungnunga, Liankhara, Tlungbuta, Lalkanglova, and Kamlova, for their part in the rising of April -May last. This decision was announced on 8 January before an assembly of chiefs.[342] Among them Rochungnunga, Lalkanglova, Lalhrima, and Tlungbuta had already been punished by the burning of their villages and destruction of their grain.[343] So were the chiefs of the Southern Lushai District such as Dokhama, Vansanga, Laltawna's mother, Laltawna, and Sailoliana. Their villages too were destroyed along with their grain while some of them were inflicted a fine of guns.[344] However, Shakespear considered these as rather mild and so the impositions of further fines. After this, on 9 January Captain Shakespear left Serchhip.[345] On the way he visited the villages west of Tlawng river, met their chiefs, and being satisfied with their conduct returned to Fort Lunglei. Davis remained in Serchhip till 23 January visiting with a small escort the remaining Haulawng villages lying within his charge and assessing them to revenue. "Everywhere I was well received" said his annual report of the North Lushai

339 *Ibid.* See also Reid, *History, op.cit.*, pp. 48-49.
340 *Ibid.*
341 *Ibid.*
342 *Ibid.*
343 FEAP. February 1893; Nos. 45-105.
344 *Ibid.*
345 *Ibid.*

Hills for the year 1892-93, "and found no difficulty in getting coolies and supplies."

But this was by no means the end of the Haulawng problem. While Shakespear was on his tour to the Northern Haulawng chiefs, Ropuiliani widow of Vandula, the chieftainess of Denlung, was directly hostile to the British. Although her late husband's brothers Lalluauva, Seipuia and Lalthangvunga were all loyal to the British, she held herself steadily aloof. All the villages under her influence were troublesome, not actively but passively obstructive. Shakespear always found difficulty in getting tribute and labour from them.[346]

In July 1892 Shakespear received information that Ropuiliani, her son Lalthuama, Dokhara and other chiefs of this group planned to attack him. He therefore arrested Ropuiliani and Lalthuama.[347] With the approval of the Government of India, they were deported to the Chittagong jail where the chieftainess died. With the capture of the obstinate chieftainess and her son, the southern Haulawng problem practically came to an end. Shakespear thus reviewed the whole operations in his report:

> The general condition of the country now, and the success we have attained in the payment of revenue and fines, seems to point to the fact that the Lushais have abandoned all ideas of combined resistance, although it is quite possible that isolated outbreaks, such as that at Jacopa, may, under similar circumstances occur for several years to come, but the force of the police on the spot, if maintained at its present strength, should be sufficient for the suppression of such disturbances.[348]

---

[346] FEBP. March 1894; Nos. 152-159, Shakespear's report concerning Ropuiliani widow of Vandula, and her son Lalthuama.

[347] *Ibid.*

[348] Reid, *Op.cit.*, p. 34

5

FINAL PACIFICATION

The settlement of the Haulawng problem was by no means the end of Lushai resistance. Throughout the winter of 1892 there were constant rumours about Kairuma's plans for an extensive rising. A.W.Davis who heard a good deal of this at Serchhip where he remained from 9 to 23 January 1893 decided to visit Kairuma's village to find out the disposition of the chief. Shakespear too had an interest in dealing with Kairuma. The southern chief Zakapa and his brother Zaduna were not only at large but were reportedly under the influence of Kairuma. Besides, the group of villages* of which Kairuma was head was located in the tract between the north and the south Lushai Hills and it was therefore thought advisable to deal with the region through combined action. Kairuma's village was only sixty-two miles from Lunglei though ninety-four from Aizawl and about the same distance from Falam in the Chin Hills.

On 26 January Davis and Shakespear met at Kairuma's village. The chief did not make his appearance but his *mantris* and people provided supplies. Though they obeyed every order given them the attitude of the villages of this group was far from friendly. All livestock from the village had been removed to the jungles. When Davis arrived there, he found the

*For genealogy of Kairuma group see Appendix No. III.

intention of resistance and said, "such action all tend to show the rumours which reached me at Serchhip, that Kairuma intended to oppose our advance, were not entirely without foundation." Had he visited the village with a small escort, Davis was quite sure, Kairuma would have attempted something of that sort. As it was "the simultaneous advance on his village of two columns of 150 men each dispelled any intention of fighting that he may even have entertained, but a bad conscience quite deterred him from venturing to meet either myself of Captain Shakespear."[349]

FINES AND IMPOSITIONS

During his tour to the Kairuma group on January 1893, Davis, for the first time, was received in a distinctly unfriendly manner. Supplies and coolies for moving the Political Officer's escort were, however, furnished with some reluctance. The chief Lalbuta, whose village lay on the route to the Kairuma group, was arrested by Davis, and confined for two days, until a fine of thirty guns was paid in for delay in furnishing supplies.[350] This incident explains the unwillingness of the chiefs of the other villages subsequently visited to meet the Political Officer, and accounted perhaps for the desertion of the villages. In the course of the same tour, Davis fined Zataia's village hundred maunds of rice for a similar delay, and, as the fine was not paid in in full five months later, he increased the fine in June 1893 to one of fifty guns.[351]

In January 1894, Davis and Shakespear again met at Kairuma's village, and their reception was as they had received in 1893, the chiefs absenting themselves but furnishing supplies and coolies for carriage only to the next village. A

---

[349] FEAP. August 1893; Nos. 175-181, Administrative report of the North Lushai Hills for the Year 1892-1893. "Tour to Kairuma group" submitted by A.W.Davis, Officiating Political Officer, North Lushai Hills.

[350] *Ibid.*

[351] *Ibid.*

halt of a fortnight was made in Zataia's village, and supplies for a force of 100 men taken from the villagers without payment, but not a single gun imposed on him last June 1893 was brought in.[352] Even coolies were not supplied except those allotted to carry rice to ration Serchhip out post. None of the coolies ordered from either Kairuma or Zataia were supplied. To enforce the order for furnishing coolies, a guard of twenty-five men was posted at Kairuma's village for some time in April, but still without effect.[353]

Alexander Porteous, the new Political Officer of North Lushai Hills in succession to Davis, continued with this policy and decided to place a guard of fifty men each in both Kairuma's and Zataia's villages to coerce them into complying with the order. Accordingly, he took an escort of 150 men to Lailen, Kairuma's village, in January 1895 and met Captain Shakespear there. The same old tactics was applied by this group. Coolies were supplied only to carry loads from one village to the next, but they refused to provide men to work in Aizawl or on the new Serchhip road then under construction. Vague promises to give labour next year were made, while all the chiefs persistently refused to present themselves before the Political Officer.[354]

To make matters worse, Kairuma abandoned his village of Lailen and not yet built his new village at Selesih across the Tuipui river. Porteous therefore, gave up his idea of placing a guard of fifty men at Lailen, as that would be futile and risky.[355] Captain Shakespear and Porteous were both of opinion that without preparation and a larger force than they had available, it would be unwise to push matters to extremities at once. They agreed to give Kairuma another chance of yielding and, in the event of his continued refusal to obey orders, to take thorough measures to bring him to submission

[352] *Ibid.*
[353] *Ibid.*
[354] *Ibid.*
[355] *Ibid.*

next cold season.[356] As a preliminary, a new fortified post was established at Selukawt, close to the village of Zataia, with the object of having a base for the collection of supplies from Aizawl. Kairuma was then ordered afresh to supply 100 coolies to work at Aizawl. This order was again ignored. Thus Porteous sent him, in the first week of April 1895, a notification that he was fined sixty guns, that if these were not given by the end of the rains, he must face the consequences.[357] Porteous had already in mind punitive measures and he was prepared to wait till 15 October, the date he fixed for Kairuma to meet the demands, before taking measures to put them into operation.

FURTHER OPERATIONS PROPOSED

Porteous had been discussing the problem of Kairuma with Shakespear from the time he took over the North Lushai Hills. Shakespear, as has been pointed out, had an interest in the matter, for with the exception of Zaduna, the whole of his charge had been pacified.[358] Both men had made it clear in their respective Annual Administration Reports for the year 1894-95 that military operations against the Kairuma group

---

356 *Ibid.*

357 *Ibid.*

358 FEAP. September 1895; Nos. 76-79, See Administration Report for the year 1894-95. In forwarding this to the Government of India, Bengal, on 8 July 1895, wrote: "The year under report was a peaceful one. The chiefs as a rule readily complied with the demand for coolies, and paid their tribute without objection. It is satisfactory to find that the payment of rice, instead of cash, is insisted on in most cases. The Lieutenant Governor considered that the obligation to supply labour in return for a reasonable rate of pay is one which is accepted by the people and consonant with their customs: it is easier to work than the system of introducing higher labour from distance countries, and it should on no account be allowed to fall into desuetude. It will be observed that although two Chief, Jaduna and his son, Kaplehya, have paid their tribute, they have not supplied coolies, and that the demand for coolies has not been insisted on, in view of the fact that, though residing within the jurisdiction of Captain Shakespear, they are really dependents of Kairuma, a chiefs under the Aijal Political Officer, who resides at the extreme point where the Lungleh, Aijal and Fort White jurisdiction meet, and has never been brought completely under the control of the British Government. A proposal has recently been made by Mr. Porteous and Captain

was absolutely essential for the final pacification of the Lushai Hills. And both of them were anxious to undertake this during the winter of 1895-96. Captain Shakespear had in fact convinced the Bengal Government of its need. The Lieutenant Governor had even told the Chief Commissioner of Assam that he was prepared to authorise Shakespear to cooperate with Assam with one hundred men if necessary.

By July 1895, after consulting Shakespear, Porteous finally decided to seek formal permission of the Chief Commissioner for military operations. The chiefs against whom action was to be undertaken were six in number with a total number of 1465 houses, viz.

Kairuma	——	455
Lungliana	——	170
Neihpuithangi	——	260
Lalbuta	——	170
Zataia	——	310
Ralthianga	——	100.[359]

In addition were the two villages of Zaduna and his son Kaphleia lying near Kairuma's, numbering between three hundred and four hundred houses. Referring to the past

---

Shakespear to send an expedition to the group of villages under Kairuma during the ensuing cold season, and the Lieutenant Governor has informed the Chief Commissioner of Assam that he will be willing to authorise Captain Shakespear to cooperate in this undertaking with a force of 100 rifles and two guns if the Chief Commissioner should consider it necessary.

The practice of making short tours with small escorts in the Lushai villages during the cold season was again adopted with satisfactory results. Captain Shakespear has not reported how far the disarmament of the Lushais has been completed, but from paragraph 16 of his report it appears that an important move has been made in this direction by the issue of orders declaring that no Lushai may posses a gun without a license, and directing the production of all the guns, which, when brought in, will be marked."

[359] *Ibid.*, Porteous to the Secretary to the Chief Commissioner of Assam, 5 July 1895.

history of the conduct of this group of villages Porteous had pointed out:

> If our authority in these hills is to be respected we can not longer delay taking action against Kairuma. The effects of the leniency with which he has been treated, have been shown in the comparative difficulty with which I this year enforced demands for labour from the Eastern Lushais, and further temporising would, I believe, seriously unsettle our position, not only with regard to the Eastern Lushais, but in the eyes of every chief in the district.[360]

He and Shakespear came to the conclusion that a force of 450 rifles would be necessary, three hundred to operate from Aizawl and 150 from Lunglei.

The general plan of operation Porteous proposed to follow was to throw into Selukawt, or Serchhip, or both, two months' supplies of all stores sufficient for a column of three hundred men and about three hundred coolies, during the months of October and November and thus dispense entirely with convoys between Aizawl and those posts during the period of operations. He intended to rely on the Lushai labour and the permanent transport corps to work during these months.[361]

He, however, left the final decision about line of advance to be adopted, whether by Selukawt or Serchhip or both, until he consult Captain G H Loch of his Military Police who was on leave. But he was quite ready in case of active resistance to bring Kairuma or his villagers to terms by destruction of grain and property, and the dispersion of Lushais wherever found in the neighbourhood of such villages and enforce disarmament.[362] Porteous believed that Kairuma would receive no outside assistance except probably from the eastern Lushais, who had recovered from McCabe's blow in 1892, and could give serious trouble again. He, therefore, proposed to

---

[360] *Ibid.*
[361] *Ibid.*
[362] *Ibid.*

increase the strength of the Aizawl garrison even though organised rising there was doubtful.[363]

The additional resources in officers, men, and transport, which he believed to be required for the proposed operations were two British Officers of standing, junior to Lieutenant Playfair, from the Military department, and temporarily attached to the Aizawl Military Police Battalion, joining not later than 25 November, and an officer to take charge of transport and general duties, and some police officers to be placed at his disposal. The ordinary strength of North Lushai Hills Military Police would be sufficient for the fighting unless casualties were numerous. If however, the eastern Lushais or the chiefs in the district would come to Kairuma's help, a reinforcement of two hundred rifles with several British Officers would be required.[364]

Regarding transport, the permanent cooly corps of 150 men, may be increased by two hundred. Of this he intended to recruit 150 in the Khasi Hills and to obtain the rest from the Deputy Commissioner of Cachar. He further suggested that the coolies would be required to arrive in Aizawl by the 20 November.[365] The cost of the expenditure, he believed, would depend on its duration, but he expected it would not be more than rupees 5100 per month in addition to the cost of local administration.[366]

Porteous' scheme also included the cooperation of a column from Falam in Burma. The Chin Hills authorities had long been concerned about the connection between some of the Chin tribes and Kairuma. Henry Newman Tuck, the Political Officer, Chin Hills, was anxious to complete the disarming of the Tlaisun and their tributaries, the Zahaus and Hualngos, in the ensuing cold weather. From the Mrung

363 *Ibid.*
364 *Ibid.*
365 *Ibid.*
366 *Ibid.*

group of villages 140 guns had been withdrawn but they were still over armed by some 150 guns. It was believed that some of these tribes were passing on their firearms to the Kairuma group. While some others were reported to have been supplied by the Lushai chiefs.[367] The Kairuma group of villages therefore assumed importance in the estimation of the Chin Hills Officers. They saw good reason for cooperating in the operations against Kairuma. Their reasons were set forth in the Tuck's supplementary report of the administration of the Chin Hills for the year 1894-95. These were:

1. Kairuma's position on their border makes his subordination of vital interest to the Chin Hills.
2. By sending a force to assist in disarming him, it would show their people that they were not taking their guns and at the same time leaving their hereditary enemies in a position to attack them with a certainty of success, and that the people will therefore more readily surrender their arms.
3. They will be able to show the Chins that their country was surrounded by territories ruled by the same nation, whose forces were always ready to help each other.
4. The subjugation of Kairuma, which would probably not be effected without the use of some little force, would be an example to the Yahows, and should have the same beneficial effect as the downfall of Thetta had in the south that year.[368]

The Chief Commissioner of Assam, William Ward, fully

---

[367] *Ibid*., H.N.Tuck to the Chief Secretary to the Chief Commissioner of Burma, No. 2821. Dated Falam, the 5 June 1895. "I believe that we can withdraw 2000 guns from these tribes, and thus render them powerless to cause us any anxiety in the future, without fighting, but every precaution should be taken to so strengthen our position at Falam that we may be able to assume the offensive with assured and immediate success."

[368] *Ibid*.

agreed with Porteous and forwarded his proposals, including the need for a column from Burma, to the Government of India.

FINAL OPERATIONS: RESULTS

From July the details of the operations were discussed and noted in the Foreign and Military Departments of the Government of India.[369] On 24 October the Commander in Chief, General Sir George White, approved of the scheme and on 10 November sanction was conveyed to Assam. The Lieutenant Governor of Bengal also expressed his willingness to co-operate in the proposed operations with a force of 100 men and two guns from Lunglei.[370] He had also addressed the Chief Commissioner of Burma with a view to obtaining his

---

[369] *Ibid.* There had been some misgivings about the strategy. The Quarter Master General had noted: "The advance of the three separate columns from different directions is considered politically advantageous, the column from Falam being classed as merely a demonstration. Mr. Porteous (his no. 240, paragraph 11) and Captain Shakespear are agreed that a force of 450 rifles are equal to any combination Kairuma and Jaduna could bring against them. This is excluding the Eastern Lushais or any other Chiefs in the districts. It is strategy to march divided and fight concentrated, but it is not so to move in weak columns liable separately to attack, though perhaps allowable where surprise is intended.

Although it may be expected that Kairuma would submit when he found troops advancing from three sides, yet if he meant to be active, both the Lungleh (100 men and 2 guns) and the Falam (100 to 150 men) columns would offer him opportunities. The columns are apparently small as more men cannot be spare from the posts. In the event of a reverse, therefore it is probable that regular troops would be called for. The proposal to place stores, with small guards, at the outposts of Sellukot and Shership, in advance of the column and nearer to Kairuma than to Aijal and Lungleh, is decidedly a temptation to Kairuma to raid the outposts, when he knows an advance is being made against him.

It appears that the Aijal column will rely on these stores during operations, and it would be while to strengthen these posts when the columns advance beyond them, and is even more necessary while they are being stored." The Adjutant General had agreed: "The proposed 'strategy' is very much open to criticism, but it will probably answer the situation.

If reinforcements are called for from the military- and I understand that Silchar is to be garrisoned by 200 extra Gurkhas for this purpose if required- then the command of the expedition should be placed in the hands of the military authorities."

[370] *Ibid.*

co-operation with a small force of 100 to 150 men from Falam.

The expeditionary force consisting of three British Officers, Captain Loch, and Lieutenant Wilson and Playfair, six Native Officers, and 300 rifles of North Lushai Hills Military Police, including the Political Officer and the Surgeon Captain Hare and Anley supported by the 347 coolies and eight *Sardars* left Aizawl Fort for the Selukawt outpost on 16 December 1895.[371] The Burma Column was accompanied by H N Tuck and commanded by Captain H E Whiffin with Lieutenants Cobbe, Tunner and Drage, and Assistant Political Officer, Fowler, with 100 Chin Hills Military Police; while the Lunglei Column consisted of 113 rifles under Major Shakespear.[372] The force from Aizawl, however, was further strengthened by one of the two mountain guns with its detachment of five men of the Lunglei force, after Porteous consulted Shakespear at Selukawt.[373] On the 22 December, the Aizawl Column left that outpost and met with the Burma and Lunglei forces at Kairuma village on the Christmas day of 1895.[374] The marches of each of the columns were unopposed by the Lushais. Thus the whole force garrisoned at Kairuma village and the chief Kairuma as usual took refuge to avoid the British forces.[375]

While in the Selukawt outpost, seventeen guns had been attained, ten from Kairuma and seven from other villages, Kairuma's *mantri* protested that they possessed no more than ten guns in all.[376] After discussion with Shakespear and Tuck, Porteous decided to demand from Kairuma eighty guns to be

---

[371] FEAP. March 1896, Nos.46-62, *Op.cit.*(No.57) Report of the Operations against the Lushai Chief Kairuma and the village of his group by .Porteous, Political Officer, North Lushai Hills to the Secretary to the Chief Commissioner of Assam

[372] *Ibid.*

[373] *Ibid.*

[374] *Ibid.*

[375] Reid, *History of the Frontier*, p. 33. See also FEAP. March 1896; Nos. 46-62, K.W. No.2. Demi Official from A.Porteous to H W G Cole

[376] FEAP. March 1896; Nos. 46-62, K.W. No.2.

brought in by 30 December 1895. The *mantri* was told of this on the 26 December.[377]

In spite of that Kairuma submitted only fifty three guns two days after he received information and did not surrender the rest up to the expiry of the fixed date.[378] Zaduna also submitted only seventeen guns out of the demand for sixty. The failure of Kairuma and Zaduna to comply with the order was obviously perceived as recalcitrance by the British Officers. Accordingly the greater part of Kairuma village was destroyed, and the parties were sent out to destroy and bring in grain and livestock. By the end of December 1895 the forces had collected as many as eighty maunds of rice, fifty pigs, and one *mithun*. Shakespear's troops who acted separately also captured thirty-one *mithuns*.[379]

Following the appearance of Kairuma's *mantri*, who said that some more guns would be surrendered, the work of destruction of grain was temporarily suspended and that functionary was told that the guns should be surrendered within three days.[380] This was just as well for the cooly convoy, sent out to bring in supplies from the Selukawt out post had not yet returned and would have been exposed to the Lushai attacks in retaliation. In all probability it was for the protection of his transport train that Porteous suspended the active operations. Still Porteous did not stop the collection of rice. At the same time, the security of the position of the Aizawl column, even though absolutely secure against any attack, was further strengthened for fear of Kairuma group.[381]

Up to the first week of January 1896, neither Kairuma nor Zaduna surrendered any of the promised guns though they did not offer any opposition to the destruction and seizure of their properties. Their inactivity led Porteous to believe

---

377 *Ibid.*
378 *Ibid.*
379 *Ibid.*
380 *Ibid*
381 *Ibid*

that he had overestimated the number of guns held by these villages.[382] Further destruction of crops which he had believed would quicken the surrender of guns was accordingly stopped.[383]

Porteous began to despair about capturing Kairuma. On the 4 January, a detachment of 100 men under Lieutenant Wilson and Playfair was sent to Lailen, the old village of Kairuma.[384] But this did not produce any effect. He had informed the latter through his *mantri* that he should come and meet him and depart freely without fear of arrest, but that if he declined to do so he should be prepared for military operations for his capture.[385] In spite of this Kairuma did not appear before the Political Officer though he supplied coolies from his village and paid the fine of thirty *mithuns*, in addition to the surrender of two thirds of the fine of guns. Porteous decided to garrison Kairuma's village with about 100 rifles. His plan were to extract the remaining fines in coolies free of charge to be supplied to the Public Works Department for work on the Serchhip road,[386] and to leave the outstanding fine of guns to be recovered gradually by the garrison.[387] Thereafter he decided to proceed towards five other villages of the group.

Kairuma was still at large, and it was impossible to obtain information of his whereabouts. It was believed that he had gone to meet Lalburha. Thus, an attempt to persuade the chief to appear before the Political Officer was not successful. The main reason for Kairuma's refusal to submit himself was that he was afraid of deportation like many other Lushai chiefs. Apart from this, Kairuma has a superstitious dread of meeting the Political Officer.[388] Since however, Porteous had

382 *Ibid.*

383 *Ibid.*

384 *Ibid.*

385 FEAP. March 1896; Nos. 46-62, K.W. No.2.

386 *Ibid.*

387 *Ibid.*

388 *Ibid.*

collected thousands of maunds of paddy and stored at the village and large amount had been destroyed, he decided to use such collected *dhans* for the supply of the forces, whom he determined to leave them at this village.

Just as the main force was leaving the village on 6 January Kairuma's *mantri* appeared and promised that 6000 unpaid coolies would be supplied,[389] but that the balance of the guns demanded was quite impossible for them to furnish owing to the dispersed state of their villages. Further, if a few months grace were allowed, they would surrender another fifteen more.[390] This was satisfactory enough for Porteous. Two thirds of the total number of guns demanded had been already received, the village burnt and a large number of livestock had been destroyed or taken.[391]

The five villages of Lalbuta, Lungliana, Ralthianga, Zataia, and their mother Neihpuithangi were still to be dealt with. Porteous had assessed their fines in the proportion of one gun for every five and half houses as was done with Kairuma. Accordingly, ten, fifteen, and thirty guns as fines were imposed upon the villages of Lalbuta, Ralthianga, and Lungliana respectively.[392] In addition each village was to supply unpaid labour.[393]

Porteous made the *mantri* surrender ten *mithuns* and accepted 150 coolies which was promptly supplied, to work at Aizawl, as preliminary towards an agreement for the submission of Kairuma. He promised to cease all punitive measures against their villages and allowed them to rebuild Selesih two miles south east of Tlaikuang.[394] Evidently, Porteous was now convinced that it was time to adopt a

---

389 *Ibid.*

390 *Ibid.*

391 FEAP. October 1896; Nos. 56-59.

392 FEAP. June 1896; Nos. 16-22, Dairy of A.Porteous, Political Officer of North Lushai Hills, for January 1896.

393 FEAP. October 1896; Nos. 56-59.

394 *Ibid.*

conciliatory policy. But when Porteous and Captain Loch reached these villages they found them totally deserted and there was none to convey the demands. Porteous remained in the village for twelve days from 13 to 24 January 1896.[395] During this period, he succeeded in extracting ten guns from Ralthianga and twenty five from Lungliana and both of them along with Lalbuta supplied coolies to work on the Serchhip road. When there was a delay in supplying the coolies a fine of ten *mithuns* were realised from them.[396] In Lalbuta's case, because he surrendered thirty guns in 1893 to A W Davis he was made to supply coolies and *mithuns* only.[397]

Nor was Porteous successful in realising from Neihpuithangi and Zataia fines of fifty guns each which he imposed upon them in addition to the small number they had previously surrendered.[398] Like most other villages theirs were totally abandoned. Indeed Zataia did not make the slightest attempt either to pay his fine or to obtain a remission by meeting the Political Officer to explain matters.[399] As he was the next in importance to Kairuma, his case was considered rather seriously.

During his visit to the chiefs between the Tuipui and Tiau Rivers Porteous found them amenable.[400] When in Vangtlang, on 5 February 1895, a hamlet of 50 house built by Ralthiauva, Zataia's brother, and where Zataia was acknowledged as chief, he obtained thirteen guns, showing that even Zataia was not completely intractable.[401] However Zataia's *mantri* was told that at least twenty-five guns should be surrendered or else the newly cut *jhums* would be burnt and whatever properties,

---

395 *Ibid.*
396 *Ibid*
397 *Ibid*
398 *Ibid*
399 *Ibid.*
400 FEAP. June 1896; Nos. 16-22.
401 FEAP. October 1896; Nos. 56-59.

livestock and paddy found in their old village would be destroyed. Zataia at once surrendered eighteen guns and once again was told that the remainder must be brought in without delay.[402] Even after five days Zataia did not make any attempt to surrender any more guns. Porteous promptly destroyed his newly cut *jhums*. *Mithuns* were seized grain destroyed. It turned out to be a severe punishment for the people.

Meanwhile Shakespear, who had received nineteen out of the sixty guns demanded from Zaduna had destroyed half of the grain of that village and succeeded in apprehending the fugitive chief Zakapa on 31 December 1895. This was followed by the capture of Zaduna from his hiding place in a *jhum* hut across the Tiau River in the Southern Chin village of Lungler together with his wife and daughter on the night of the 4 January 1896.[403] Regarding the surrender of guns Shakespear did not receive any more from Zaduna.[404] However, the entire collection of the grain from Zaduna's group amounted up to 1500 maunds. After this Shakespear suspended operations from the middle of January.[405] Since there was little chance for the surrender of more guns from this group and since the prominent chiefs, who could control the disintegrated Lushai society, were already under restraint, further punitive measures were considered unnecessary. Therefore Shakespear left for Lunglei leaving a small force to enforce payment of the remaining guns.[406]

So it went on, guns and livestock seized and villages put to flames. After this Porteous turned his attention towards Lalburha's new village on 13 February1896 with an escort of 150 men.[407] Lalburha, still a fugitive was permitted by Davis

---

402 *Ibid.*

403 FEAP. October 1896; Nos. 56-59, A.Porteous' report of the Operations against the Lushai Chief Kairuma and the village of his group. 6 May 1896.

404 *Ibid.*

405 *Ibid.*

406 *Ibid.*

407 *Ibid.*

in 1894 to make a new village on payment as fine of fifty guns and five *mithuns*. Porteous renewed the offer on seeing that Lalburha had actually built the village. Porteous arrived after two days march from Khawhai, but found it deserted. Lalthangvunga, a near relative of Lalburha, was sent to the fugitive chief to convince him to submit or else he would destroy his village and crops.[408]

While awaiting the result of this mission, Porteous surveyed the valley of Champhai whether it will serve the purpose of an administration centre of the whole Chin-Lushai land.[409] On his return, on the 24 February, he came to know that Lalburha was disposed to submit and as a sign of his willingness had sent in an instalment of fourteen guns as his fine and promised to deliver a valuable elephant's tusk in lieu of the five *mithuns*.[410] However, Lalburha was not easily convinced to come into camp, and with great difficulty, he was induced to come within half a mile distance of it on 26 February. Porteous himself went out and met Lalburha in the *jhum*. At that interview Lalburha promised that in future he would fully obey Government Orders, and supply labour when demanded. As for the guns, he said he would not be able to fulfil the demand. He then handed over thirteen guns on the spot and sent in one more gun and the promised elephant tusk. Porteous told him that from then on he would be recognised as a chief.[411] In return Lalburha sent to Aizawl a number of coolies as a token of his submission. After the settlement with Lalburha, Porteous and Captain Loch returned to Aizawl on the 11 March 1896.[412] Soon after Kairuma too made overtures for surrender, sending his nephew to meet Porteous at Aizawl. The last great chief to

---

408 *Ibid.*
409 *Ibid.*
410 *Ibid.*
411 *Ibid.*
412 *Ibid.*

surrender asked that he might be allowed to live at peace in one of the out-hamlets of his village and promised to meet the Political Officer and gradually pay up the balance of his fines in guns. With Porteous' acceptance of this Kairuma's resistance came to an end.

The Burma column which had been operating in co-operation with Shakespear left for Falam on 5 January 1896, with forty *mithuns* captured from Kairuma's and Zaduna's groups. These were used to buy off the areas between Tiau and Tuipui rivers, which had long been claimed by the Falam chief by right of conquest. The claim was also recognised by Kairuma group. Tuck handed over the *mithuns* to Falam chief.

The operations against Kairuma and his allied group was successful in achieving the main aim of subjugation of one of the most powerful combinations of the Lushai chiefs.[413] Porteous recognised the contribution of the Burma column to the success of the pacification process:

> The prime cause of Kairuma's collapse has unquestionably been Mr.Tuck's successful disarmament this season of the Tashons and Jahows, on whom Kairuma undoubtedly leaned, but the presence of Falam force must have driven home to the minds of Kairuma and his subordinate chiefs the truth of their real insignificance, and of the essential unity of the administration exercised from the three centres- Falam, Lungleh and Aijal, respectively, in a way that no less concrete exhibition of force could possibly have accomplished.[414]

In summarising the results of these operations Porteous concluded:

> With the close of the operations against Kairuma, it may, I think, be safely prophesied that the long series of Lushai expeditions has now ended. The immediate results of the expedition have been to completely break the power and

---

[413] *Ibid.*

[414] FEAP. June 1896; Nos.16-22. Dairy of A Porteous, Political Officer, North Lushai Hills, for January 1896.

> prestige of Kairuma, and to dispel effectually the idea that any Lushai Chief...can safely ignore the orders of the Political Officer.[415]

DISARMAMENT

The most striking feature of the pacification process was the attempt to disarm the people. Since total disarmament was not found practicable, arms limitation became the objective. It will be recalled that in early 1889, an assessment was made of the possession of the firearms among the Northern Lushais and as many as seventeen hundred guns were estimated.[416] This was of course before the resistance and was considered to be a fairly accurate figure as the Lushais then had no reason to conceal guns in their possession. At the time of the uprising it was assumed that the number might have increased by around two hundred.[417] Out of these approximately three hundred were believed to have been sold to the Chins, and thus the total number of firearms with the Northern Lushais could not have been less than sixteen hundred.[418] These had been distributed among the western, eastern, and the Kairuma group of villages.[419] Among the last the number of guns was estimated to be at the rate of one gun per four houses.

To ensure the subjugation of the people and so that they abandoned all ideas of resistance, collection of firearms began from the first operations. McCabe was successful in extracting about one hundred thirty guns from the western Lushais,[420]

---

[415] FEAP. October 1896; Nos. 56-59.

[416] FEAP. August 1893; Nos. 175-181, A.W.Davis's annual report of the Northern Lushai Hills for the year 1892-93.

[417] *Ibid*

[418] *Ibid*

[419] *Ibid.*

[420] FEAP. January 1892; Nos. 51-61, McCabe's report on the Northern Lushai Hills for the year 1890-91, to the Secretary to the Chief Commissioner of Assam. The table shows the guns extracted from the western Lushai chiefs:

Sl.No.	Name of the Chiefs.	Number of guns demanded.
1	Kalkhama	40 guns
2	Lianphunga	19 guns

and this marked the beginning of the formal operations of disarmament in the Lushai Hills. The suppression of the rising of the eastern Lushais was accompanied by the extraction of about two hundred guns.[421] This ensured that the Lushais in the north were more or less disarmed.

A W Davis, Political Officer of the North Lushai Hills successfully extracted an equal number of guns from the Haulawng group during the operations in the winter of 1893. Here too the people were thoroughly disarmed.[422] Davis was aware that it was not totally possible to get hold of all the guns in the possession of the Lushai people. The guns were old flint lock muskets and Davis knew that while the Lushais could manufacture powder they were dependent on outside sources for the flint. These generally came from the plains people through the *bazars* at Changsil, Tuirial, Tipaimukh, Demagri. Davis thereafter had attempted at strict watch over these *bazars*.[423]

His successor, Porteous successfully extracted a number of guns from the Kairuma villages as punitive fines. However, the number of guns was somewhat overestimated.[424] These villages were believed to be the most self-sufficient in firearms and were known to have sold very little to the Chins as

3	Thanghulha	10 guns
4	Sailianpuia	25 guns
5	Lalhrima	16 guns
6	Hrangkhupa	10 guns
7	Lalsavuta	10 guns
8	Hmingthanga	10 guns
9	Thatliana	5 guns
10	Khama	7 guns
11	Thawmpawnga	Not recorded
12	Tulera	2 guns
13	Liankunga	30 guns
	TOTAL	184 guns.

421 *Ibid.*

422 *Ibid.*

423 FEAP. March 1896; Nos. 46-62, Demi-Official letter of A.Porteous.

424 *Ibid.*

compared to the eastern and western Lushais. This was because they were always at feud with the Pawis to the south-east of their boundary. As such possession of a good number of firearms was essential for their own defence. The following table shows the demands and the actual realisation of firearms from Kairuma and the villages of his group.

Name of chiefs.	Number of houses	Guns demanded	Actual surrender	Balance
Kairuma	455	80	53	27
Neihpuithangi	260	50	12	38
Zataia	130	50	25	25
Lalbuta	200	40	40	Nil
Ralthianga	100	15	15	Nil
Lungliana	170	30	26	4
Total	1315	265	171	9

This clearly illustrates that only two third of the total arms possession had been extracted.[425]

Major Shakespear was also equally successful in his operations against Zaduna group, and by the middle of January 1896, he received forty-five guns surrendered in all out of the sixty guns demanded.[426] Three-fourths of the total demand had been surrendered, and it was equally clear this group was fairly disarmed.

In describing the general position of the Chin-Lushai country the Foreign Secretary to the Government of India said:

> The Chins and the Lushais are not formidable fighting races; they have little or no cohesion or military organisation, and, as an enemy, were troublesome only

[425] FEAP. October 1896; Nos. 56-59, Report of A. Porteous, Political Officer of the North Lushai Hills, to the Secretary to the Chief Commissioner of Assam, 6 May, 1896.

[426] *Ibid.*

> through the difficulty of their country, our ignorance of it, and the facilities it afforded for ambuscade and for guerrilla warfare. The country is now known; the tribes have had evidence of our power and of the fact that they are surrounded; they were at the outset very poor armed and have since had to surrender most of the fire-arms they possessed. The total number of fire arms withdrawn to date exceeds 8000, nearly 7000 had been withdrawn from the Chins, and 700 from the Northern Lushai hills.[427]

After the establishment of the Lushai Hills District, the Government of Assam, advised by local officers introduced a system of licensing guns, under the Arms Act XI of 1878.[428] Captain Shakespear introduced this in 1898 when orders were issued that no Lushai may possess a gun without a license. He further directed that all the guns were to be brought in to be marked.[429] There was a favourable response to this from the people. Marking of guns was certainly an important achievement: the Lushais were happy to have their guns licensed and the Superintendent was able to keep a tract of the number available in his District. Punishment for unlicensed guns made the Lushais more responsive. Shakespear was opposed to send punitive detachments to villages suspected of having some unlicensed fire-arms. Certainly the amount of misery inflicted on innocent people by this method would give rise to a sense of insecurity and discontent, which Shakespear felt was far more dangerous than the possession of a few guns. At any rate in the future he was sure to hear of unlicensed guns in time.[430]

---

[427] FEAP. October 1896; Nos. 28-35, Secretary, Government of India, to the Chief Secretary to the Government of Bengal. 8 September 1896.

[428] FEAP. February 1898; Nos. 104-142, Assam Chief Commissioner's proposal submitted to the Foreign Department. 17 July 1897.

[429] FEAP. September 1895; Nos. 76-79, H J S Cotton, Chief Secretary to the Government of Bengal to the Secretary to the Government of India, Foreign Department. 8th July 1895.

[430] *Ibid.*

Disarmament more than anything else contributed to the final pacification of the hills. Shakespear in his administrative report of the Lushai Hills for the year 1898-1899 sums up the results of his efforts:

> There are now in the Aijal Sub-division, 571 licensed guns, of which 32 are held by *sepoys* and other foreigners, leaving 539 in the hands of the Lushais. In the Lungleh Sub-division there are 273 licensed guns. The total for the district is therefore 812 as against about 1040, which might be licensed at the scale sanctioned by the Chief Commissioner, viz. 1 gun in 15 houses.[431]

With regard to the possession of the legal and licensed guns of the Lushais, Shakespear further reported in a separate paper, which was proudly forwarded by the Secretary to the Chief Commissioner of Assam to the Government of India, thus:

> The total number of licensed guns in the district is now 812, which is somewhat below the sanctioned scale of one gun to fifteen houses. The established policy in these hills is to license guns with caution and moderation, imposing a maximum limit as far as possible, and to confiscate all unlicensed guns, fining heavily the owners of such guns and liberally rewarding informers. This policy has stood the test of time, and a systematic control is now exercised. No complete disarmament has ever been attempted, and the Chief Commissioner does not recommend any such measure.[432]

---

[431] MSR. Pol-6/ CB-1, Shakespear's Administration Report of the Lushai Hills for the Year 1899.

[432] FEAP. October 1899; Nos. 35-41, Annual Administrative Report of the Lushai Hills for the year 1898-1899, forwarded by Secretary to the Chief Commissioner of Assam to the Government of India Foreign Department. 12th June 1899.

6

TOWARDS CONSOLIDATION

With the final pacification of the Lushai hills the British were able to turn their attention to the long delayed question of the administration of the Chin-Lushai land. The administrative control of these hills by three different authorities, in the north by Assam, to the south by Bengal and the Chin Hills in the east by Burma, was a temporary measure and it was as inconvenient as it was expensive. The existing arrangements, under which these newly incorporated tracts were held bv the three governments were far from satisfactory in so far as its administration was concerned. It had been recognised from some time past that this "tripartite division of authority" was objectionable and ought to be simplified.

The financial implications of the problem too was equally realised by the authorities. In July 1891 the Government of India, when recommending to the Secretary of State for sanction to the heavy outlay of over five and half lakhs of rupees proposed by the Bengal Government for the administration of South Lushai Hills had observed: "It is probable that ultimately it may be found possible and desirable to consolidate under one administration the whole or the greater part of the territory in the occupation of the various tribes now separately controlled from Bengal, Burma and

Assam."[433] The Secretary of State for India had agreed that the cost of administration of the South Lushai hills alone was too expensive: "The cost is seemingly disproportionate to the area and population..."[434] Almost all the local Officers were concerned with the expenditure incurred upon the administration and the difficulties in transport and communications.[435]

The need for a single authority over the whole Chin-Lushai country apart from cutting down cost was seen as desirable from the very beginning. Captain Shakespear, when Superintendent of the South Lushai Hills, had thus pointed out the practical inconvenience of administering a single ethnic and political entity by three different authorities when he related one incident:

> Tingtua asked me several times to allow him to attack Vongthu. He says Vongthu has attacked him before, and their recent attempt proves they are not inclined to be friendly in spite of their submission to General Symonds, and so he wanted to attack them. Of course I refused to allow him, but at the same time I must say it is a little hard on our people if we stop them settling matters in their own way, and yet the village nominally under Haka are allowed to send out raiding parties whenever they like.[436]

The uncertainty of control over the areas between the Kolodyne River and Falam was further pointed out by Shakespear:

---

[433] FEAP. September 1891; No. 56.

[434] FEAP. September 1891; No. 9.

[435] FEAP. October 1896; Nos. 28-35, C W Bolton, Chief Secretary of Bengal to Secretary, Government of India, 4 May 1896. The cost of administration of the whole of Chin-Lushai land can be seen in the report by C W Bolton Bengal's Chief Secretary, as shown below:

North Lushai Hills	1893-'94 to 1896-'97	Rs. 5,46,300.00
South Lushai Hills	1893-'94 to 1896-'97	Rs. 4.94,000.00
The Chin Hills	1893-'94 to 1896-'97	Rs. 2,56,500.00
	Total =	Rs. 12,96,800.00

[436] FEAP. September 1892; No. 36, Chin-Lushai Conference, Note by C A Elloit, 28 January 1892.

> The two systems cannot work side by side. It is evidenced that peace and order cannot exist west of Kolodyne, if east of that river the old state of things is allowed to continue.[437]

David Robert Lyall, the Commissioner of Chittagong Division, was of the same view with Shakespear. In forwarding the report of the Superintendent of South Lushai Hills, to the Lieutenant Governor on 5 November 1890, he wrote: "It is impossible that good government of the new territory should go on, if different degrees of influence are exercised by Bengal, Burma, and Assam."[438] The military authorities too were keen to dispose the question of the future administration of the Chin-Lushai country as the military garrison in eastern India and Burma had to be settled.[439] They accordingly endorsed the idea of the conference. Out of the practical inconvenience with regard to the expenditure of each of the territory, the Secretary of State for India found it necessary to consolidate the Chin-Lushai land under one administration and made a proposal for a meeting of the Officers of these three governments to work out the possibility of simplifying the existing system and subject it to the control of one administration.[440]

EARLY RECOMMENDATIONS

A conference was thus convened at Calcutta on 25 January 1892 known as the Chin-Lushai Conference, "to discuss civil and military affairs connected with the control of the Lushai and Chin Hills". The conference was chaired by Sir Charles Alfred Elliott, Lieutenant Governor of Bengal and was attended by Lieutenant General James Dormer, Commander in Chief, Madras, Sir Alexander Mackenzie, Chief Commissioner of Burma, William E Ward, Chief Commissioner

---

437 *Ibid.*

438 *Ibid.*

439 *Ibid.* See note by E H H Collen, "The Military Situation in Eastern India and Burma," 4 September 1891.

440 FEAP. September 1892; Nos. 9-62.

of Assam, Sir Henry Mortimer Durand, Secretary to the Government of India, Foreign Department, Major General James Browne, Quarter Master General of India.[441]

For four days the conference discussed various aspects of the future administration and military control of the Chin-Lushai Hills. On the important question, "whether the practical inconvenience of the present system of administration is as great as has been represented, and whether it is desirable to make a change and what remedies would be practicable,"[442] Sir Alexander Mackenzie wrote a minute on the 27th which effectively sealed the union of the Chin Hills with the Lushai Hills. His view was that putting the Chins under Assam would be impracticable for some years to come, and its desirability never.[443] He saw a good deal of difference between the Lushais and the Chins:

> ...it is a mistake to treat the tribes inhabiting these hills as though they were one and a same people. Ethnologically they may be so, but politically they are a congeries of independent, and even hostile communities, looking out of their hills towards the plains from which they severally draw their supplies of salt and (hitherto) of slaves. The Chins on the east look to upper Burma; those on the west to Sandoway, Kyaupyu, and Akyab. In all these three district we have Chin communities under settled administration and paying regular taxes, as well as a fringe of hills Chins who are more loosely managed.[444]

---

[441] *Ibid.*, No. 31.

[442] *Ibid.*

[443] *Ibid.*, Demi Official Letter from A Mackenzie to the Viceroy of India.

[444] *Ibid.* Mackenzie drew this idea from Macnabb and Carey who supported this idea and furnished through briefing to Mackenzie prior to the conference. Carey upheld that the Chins have nothing in common with the Lushais, and that their historical connection, tribal sympathies and political interest were with Burma. Both Carey and Macnabb discouraged the plan with regard to the difficulties which may arise in constructing roads and telegraphic communications. Macnabb held the idea that the combination of the three districts would be too vast for one man to look after.

This was immediately contested by Sir Charles Elliott: "All the savages, whether called Chins, Lushais, or other names, who have never come under the British control before... should be under one Officer,"[445] had always been his view. Refuting Mackenzie that there was no inconvenience from the divided control of these hills, he stated:

> It seems clearly established that raids from tribes under Haka upon tribes under Lungleh are of frequent occurrence, and have occurred quite recently; that the savages are thoroughly alive to the advantage of playing off one administrator against another; and that my most experienced officers are strongly convinced of the necessity of uniting all authority in Chin-Lushai land under single head.[446]

Both maintained their respective positions. On the 29th at the final meeting the Conference adopted the majority view that:

1. the whole tract of country known as the Chin-Lushai hills should be brought under one Administrative head as soon as this can be done and that it be placed under the Chief Commissioner of Assam.
2. this step should not be taken immediately. As matters then stood, the difficulties of communication, of supplies and of transport were very serious, and it would in any case be necessary to suspend action until after the close of the present cold season's operations in the Chin and Lushai Hills.

   Sir James Dormer and Sir Alexander Mackenzie were however not prepared to accept as final any decision as regards the eastern part of the tract i.e. Chin Hills till further information was obtained.

---

[445] FEAP September 1892; Nos. 30-40, Question raised in the military department for members of the Chin-Lushai Conference to consider and their reply.

[446] *Ibid.*, note by Sir Charles Elliott, 28 January 1892.

3. the first thing to be done for the control of this tract was to improve communications between the important places such as Cachar and Aizawl, Aizawl and Lunglei, Aizawl and Manipur, and the posts situated respectively on the eastern and western side of the tract. The opening out of these lines was a work of pressing importance. The necessary commissariat start should also be provided to arrange for transport and supplies, till the tract was able to provide them for itself.
4. The boundaries of the new administrative area should be, generally speaking, the boundaries of the tract occupied by the tribes newly brought under British control, but the details of those boundaries could only be settled after consultation with local officers.
5. North and South Lushai Hills, with such portions of the Arakan Hill Tracts should be placed under Assam at once on condition that - (1) complete transport and commissariat equipment for supplies from Chittagong to South Lushai, and from Cachar to North Lushai, being provided; (2) funds granted for road and telegraph from Aizawl to Lunglei.

These proposals in their turn were discussed at length in the Viceroy's Council. Lord Lansdowne then noted:

> I am disposed to favour the view which I understand to have been held by the majority at the Conference, viz, that as a question of principle the whole of these hill tract ought to be under one management, and governed from inside by a single specially selected officer, instead by several officers belonging to different local administration.[447]

---

[447] *Ibid.*, Note by Lord Lansdowne, 3 June 1892. Lansdowne further added: "It has always seemed to me possible that eventually the new district thus created might include everything to the west of the Myittha valley and of the valley

In view of Mackenzie's opposition the Chin Hills was allowed to remain with Burma. With this exception it was accordingly decided by the Governor General in Council that the whole of the Lushai country should be brought under the Chief Commissioner of Assam, and the transfer of the South Lushai Hills from Bengal to Assam should be made as early as possible. The Chittagong district was also to be transferred to Assam.[448] If it was found that there were certain difficulties in the way of immediate transfer of the district the Hill Tracts should be transferred in advance of the rest. The Bengal and Assam administration was to be consulted as to the expediency of transferring the whole Chittagong division to Assam, and the Burma and Assam administrations as to the future position of the boundary between them in the North Arakan Hill Tracts. An estimate of the cost of transfer was ordered be submitted to the Government of India.[449]

---

which runs northwards from it towards the frontier of Manipur. But neither the Conference, nor any one else, has proposed that so extensive a change should be made at once, and the only question, I understand, is whether the Government of India should for the present remain content with the comparatively modest rearrangement (now under discussion with the local Government) which will give the north and south Lushais and part of the Aracan Hills to Assam, or whether we should commit ourselves definitely at a preference for the policy recommended by the majority. No declaration which we can make upon this point can bind the Government of India, but I should myself prefer to place it upon record that whatever steps are now taken, are taken in the believe that they lead in the direction to which I have pointed above, and that, when communication have been improved, we look forward to a larger and more thorough measure of consolidation."

[448] FEAP. July 1898; Nos. 40-62. Proclamation of the transfer of the South Lushai Hills under the administration of Assam.

[449] *Ibid.*, Extract from the proceedings of the Government of India, Foreign Department. H M Durand, Secretary to the Government of India, clearly reflected the recommendation along with the failure of the amalgamation of the three district under on administrative head, thus: "The majority of the conference are of opinion that it is very desirable that the whole tract of country known as the Chin-Lushai Hills should be brought under one administrative head as soon as this can be done. They also consider it advisable that the new administration should be subordinate to the Chief Commissioner of Assam. Sir J Dormer and Sir Alexander Mackenzie would differ any final decision as regards the eastern part of the tract till further information is obtained."

Bengal Government accepted these directions but the Chief Commissioner of Assam demurred, principally on financial grounds. On being assured liberal financial grants by the Government of India, viz. provincialisation of the Lushai administrative cost and deduction of the cost of administration of South Lushai hills from Bengal and its transference to Assam, construction of the Aizawl-Lunglei road as an imperial charge, and organisation of supply and transport system by the Military Department, the Chief Commissioner of Assam accepted the proposal on the 19 March 1893.[450] By then Lushai resistance and the pacification process was at its peak and the whole question had to be deferred. The three Governments continued to administer their own tracts and conduct military operations, all independent of one another.

AMALGAMATION OF NORTH-SOUTH HILLS: PROPOSAL REVIVED

Some four years later, with a change in the situation, improvement of communications and the general consolidation of the British rule in the hills, the Government of India assumed that the time had come when very substantial reductions in expenditure could be undertaken and a single administration established. The average expenditure on the Chin-Lushai land during the years 1893-1894 to 1896-1897 had come up to Rs.12,96,800/-each.[451] It was obvious that it was more expensive to administer these hills by three different authorities than it would have been by a single administration. Moreover, the Government of India clearly realised the fact that the Chins and Lushais were not formidable fighting races: they had little or no cohesion or military organisation, and, as an enemy were troublesome only through the difficulty of their country.

---

[450] Suhas Chatterjee, *Mizoram Under British Rule,* (Delhi, 1985), pp. 125-126.

[451] FEAP. October 1896; Nos. 28-35.

The Government of India, therefore, suggested a conference of the local officers to work at the details of the future administration of the tracts.[452] A conference of the officers of the two Lushais and the Chin Hills was accordingly held between 14 and 18 December 1896 at Lunglei. It was attended by Alexander Porteous, Political Officer, North Lushai Hills; Robert Henry Sneyd Hutchinson, Officiating Superintendent, South Lushai Hills; Henry Newman Tuck, Political Officer, Chin Hills; and Captain G H Loch, Commandant, North Lushai Military Police.[453]

The conference discussed and agreed upon a variety of issues, the principal ones being:

1) Concentration of Military Police Battalion in certain central position and withdrawal of isolated and expensive outposts;
2) Reduction of forces and establishment to the minimum necessary for the maintenance of such direct political control as may still be required;
3) Establishment of good communications from north to south both in the Chin Hills and Lushai Hills, and from east to west between them;
4) Enforcement of tribal village responsibilities, allowances being, if necessary, given to headmen;
5) Amalgamation of North and South Lushai Hills.[454]

It will be seen that the question of the unification of the whole tract by including the Chin Hills did not form part of the discussions of the conference. But the amalgamation of north and south Lushai Hills was taken as already decided and the conference merely reiterated the transfer of the south

---

[452] Sir Robert Reid, *History of the Frontier Areas Bordering on Assam*, (Shillong, 1942), p.39.

[453] *Ibid.*

[454] FEAP. October 1896; Nos. 28-35, From the Secretary to the Government of India, to the Chief Secretary to the Government of Bengal. 4 May 1896.

to Assam on both the political and financial grounds. The Government of India when receiving the proceedings of the conference was happy to note that it would effect an annual saving of Rupees two lakhs.[455]

The Chief Commissioner of Assam when informed once again agreed to taking over the South Lushai Hills, but desired that the change should be made after some months to allow him time for the settlement of various questions that needed to be disposed of. He had suggested that 1 October 1897 would be a suitable date for transfer to which Bengal's Lieutenant Governor agreed.[456] Ward further stated that Lunglei should continue to be provisioned from Chittagong and that the Government of Bengal should give all the necessary assistance in these regard. Assam's position was thus explained:

> I can see in brief that there are many objections to taking over the charge of the South Lushai Hills by this administration. On the other hand, I clearly perceived the advantage of amalgamation, and admit that they preponderate over the drawbacks. It is not possible to maintain that uniform and concerted action and persistent policy, which are so necessary over this large and recently annexed territory. When its affairs are controlled by separate and independent governments, I have no doubt also that amalgamation will be followed by improved efficiency in administration and by large reduction in expenditure.[457]

The station of Demagri posed a small problem. It was not, in fact, situated within the area of the Southern Lushai Hills, but was geographically a part of the Chittagong Hill Tracts.[458] Its inhabitants were the descendants of Rothangpuia

---

455 *Ibid.*

456 FEBP. February 1898; Nos. 104-142, C.W. Bolton, Chief Secretary to the Government of Bengal, to the Secretary to the Government of India, Foreign Department. No.391-P. 8 February 1897.

457 *Ibid.*, Note by the Chief Commissioner of Assam on the proposed transfer of the South Lushai Hills from Bengal to Assam.

458 *Ibid.*

and of the Lushai clans. In 1892 Charles Elliot, Lieutenant Governor of Bengal, had ordered that for administrative purposes, Demagri should be considered a part of the South Lushai Hills.[459] Its importance lay in the fact that it was the base of all communications to the Southern Lushai Hills. The Commissioner of Chittagong therefore considered it absolutely necessary that Demagri should be under Assam in the future administration of the area and that Sir Charles Elliott's arrangement should continue.[460] This was agreed upon and Demagiri was to remain the entry point to the South Lushai Hills and the terminus of trade from the hills.

LUSHAI HILLS DISTRICT: THE CHIEFS

On the 27 January 1898, the Government of India accorded its approval to the proposed transfer of South Lushai Hills to Assam and thus the whole of North and South Lushai Hills constituted one charge under the administration of Assam.[461] The district was placed under the charge of an Officer designated as Superintendent with headquarters at Aizawl. The term "Superintendent" was considered as a more suitable designation, because 'Political Officer' used in the

---

[459] Reid, *op.cit.* p. 40.

[460] B Lalthangliana, *Mizo Chanchin*, (Aizawl , 2001), p.478.

[461] FEAP. July 1898; Nos.40-62, The Government of India's Proclamation of 1 April 1898 thus stated: Whereas the territories known as the South Lushai Hills, where a proclamation No. 1697-E., dated 6 September 1895, issued under the Government of India Act, 1865 (28 and 29 Vict., Cap. 17) Section 4, included within the lower province of Bengal, and whereas the Tract known as Rothangpuia's villages, including Demagri, in the Hill Tract of Chittagong is also included within the said lower provinces, and whereas it is expedient that the said territories and tracts should now be place under the administration of the Chief Commissioner of Assam; know all men, and it is hereby proclaimed, that the Governor General in Council has been pleased, in exercise of the power conferred by Section 3 of the Government of India Act, 1854 (17 and 18 Vict., Cap. 77) and with the sanction and approbation of the Secretary of State for India, to take the said territories and tract under his immediate authority and management, and to place them under the administration of the Chief Commissioner of Assam, and further to direct that henceforth they shall be included within the province of Assam.

North Lushai Hills was not appropriate as the duties of a Superintendent were quite different from those of Political Officers employed under the Foreign Department. The term "Deputy Commissioner" was equally unsuitable, as it failed to mark the distinction. Thus, "Superintendent" was accepted for the designation of this Officer, without making any change in his status or allowances as a member of Assam Commission.[462] The Government of Assam by proclamation, on the same date as that of the Government of India, appointed Major John Shakespear as Superintendent.[463] The whole district was divided into two sub-divisions, the North with its headquarters at Aizawl and the South at Lunglei under the charge of a Sub-Divisional Officer.

Prior to the formal acquisition of the Lushai Hills, strengthening the Lushai chiefs for the administration of the Hills had envisaged by the British Officers. It may be recalled that David Robert Lyall, the Commissioner of Chittagong Division, was one of the first officers to realise the importance of the chiefs in the administration of the hills. In a note on the future management of the western portion of the country between Chittagong and Burma he had remarked that:

> The nature of the country and of the people is such that for the present at least any attempt at government in detail would be expensive, and our knowledge of the people and custom is very small. I would therefore recommend that for the present the system of government through chiefs should be fully recognised...For the present I think we should recognise the existing state of fact and govern through chiefs, recognising a chief for each village, and not recognising any one chief as paramount over any tract of country, unless compelled to do so hereafter by the force of circumstances... I think it would be a mistake to attempt to govern too minutely or to attempt anything like district government...[464]

---

462 *Ibid.*

463 *Ibid.*

464 FEAP. August 1890; Nos. 238- 262, D R Lyall, Commissioner of Chittagong Division to the Chief Secretary to the Government of Bengal.

Lyall had clearly visualised the position of the Lushai chiefs not only when the process of pacification was on but for another generation at least when *Pax Britannica* would have been fully established throughout the district. What the chiefs should be required to do in the future administration of the Lushai hills was thus suggested.

1. Each village and chief should be made responsible for maintaining, improving and, if so ordered, the making of such roads round his village as the Superintendent may order him to. The labour should be paid for thus enabling the men to pay their tax as proposed hereafter. The Superintendent should have power of fining any chief not keeping up his road and of compelling him and his people to do the work by force.
2. Each chief should be made responsible for the collection and payment of the tax of his village.
3. There should be a meeting of chiefs each year at the central post, and attendance should as far as possible be compulsory, as evidence of their acknowledgement of British sovereignty, and absence should be punished by fine.[465]

But Lyall made it clear that the Government should not interfere with the village administration in civil and social matters, but concern themselves in the preservation of the public peace. All internal matters were to be left to the chiefs.[466] To enforce Government orders effectively, he had added that the military force should be one force under one head, and not under three governments as it then was.

Lyall's proposals became the basis of the British administration of the Lushai hills and successive district Officers did all they could to strengthen it. A W Davis the Political Officer of the North Lushai Hills, thus explained to

---

465 *Ibid.*
466 *Ibid.*

his successor Alexander Porteous when handing over of the charge to him in 1894:

> I always held the chiefs of villages responsible for the behaviour of the people, and upheld their authority to the best of my ability. I have repeatedly told them that this policy will be consistently followed, and that, as long as they behave themselves as they should, their orders will not be interfered with, even though the order may appear to us at times a little high handed, and not quite in accord with abstract ideas of justice. In this connection, it is well to remember that no chief can very greatly misuse his power or oppress his people. Were he to do so, his village, and with it his own importance, would quickly diminish, as the people would migrate to other villages. In upholding the authority of the chiefs, I have, as a rule decline to take up appeals against their orders on petty cases as it only diminishes a man's authority...[467]

John Shakespear when holding charge of the South Lushai Hills too had expressed the same views and recommended the delegation of tribal and village responsibility to the chiefs. "I am convinced" he had stated, "that it is better to uphold the government of the chiefs and to govern through them. With this view, I have submitted proposal for educating the sons of the chiefs."[468]

The Lieutenant Governor of Bengal, while considering the reduction of the expenditure in the Lushai Hills administration also remarked that the Lushai people should be left to govern themselves according to their own custom subject to the Political Officer's advise and supervision.[469] Likewise, Sir Henry Cotton, the Chief Commissioner of Assam, stated in 1897:

> The policy of the Political Officer and of his Assistants should be to abstain from avoidable interference with existing Lushai traditions and customs, but while

[467] Reid, *op.cit.* p. 41.
[468] FEAP. October 1896; Nos. 28-35.
[469] *Ibid.*

> maintaining the patriarchal powers exercised by the hereditary chiefs to insist at the same time on implicit compliance with their own orders in respect of really important matters. Thus the maintenance of the supremacy of British rule is quite consistent with this policy of working through the chiefs, and not independently of them and with the fullest recognition of the authority of the chiefs in their own village and over their own village."[470]

What needs to be emphasised, however, is that Lyall's ideas became strengthened with the experience of the protracted resistance from 1890 to 1896. It was clear that the chiefs had a tremendous hold upon the people, who had followed them in their obviously one sided struggle against the British. It was equally clear that relation between chiefs and people were governed by custom and tradition, as were the latter's power and privilege. No chief could make impositions and demands which custom and usage did not allow him. Thus the pattern of administration in the Lushai Hills was of a minimum interference, respecting the tribal laws and customs, and allowing the chiefs to exercise their authority on local matters,[471] with the full backing of the British Government. This was the indirect rule that was established in the Lushai hills.[472]

The Chief Commissioner of Assam by proclamation provided the rules for the administration of the Lushai Hills, rules which with periodical modifications were to remain in force till the end of British rule. Its main principle was the

---

470 ASR Foreign 'A' July 1897, Nos. 28-31.

471 Lalrimawia, "A Cultural Development Profile In Mizoram", in *Pialral*, A Historical Journal of Mizoram by Mizo History Association, (Aizawl, 1995), p. 43.

472 This kind of administration through the indigenous Chiefs or "Indirect rule", had already been introduced in the Khasi hills where the chiefs or *Syiems* were retained and villages were governed through them. In the Angami Naga hills, no such institution existed, the British created the *gaon burhas* or villages heads, who along with the *dobashis* became the link between the people and the district administration..

empowerment of the Lushai *Lal* or the chief for the internal control of their own village.[473] The Lushai '*Lal*' was to continue to rule with the help of '*Upas*' or Elders and his powers, responsibilities and privileges were to be the same as defined by custom and usage. Unlike the problem of a single administrative authority over Chin-Lushai country which took eight years to dispose off the broad principle of the administrative system was worked out even as the process of annexation was underway.

As the indigenous system of village administration by the chiefs was to continue, it was necessary to assign them powers and functions. Apart from the general administration of his jurisdiction, the chiefs' responsibility was the upkeep of the village paths, water points, conduct occasional meetings of the '*Upas*', framing of general village policy and of activities to promote '*Tlawmngaihna*',[474] the maintenance of accounts of all fines imposed, records of deaths and births and of proceedings.[475] They were to keep Superintendent informed of any distress or difficulties under which his villagers were suffering. The chiefs were responsible for the behaviour of their subjects. Besides these, was the important function of carrying out the responsibility for collection of house tax and supply of labour.

The chiefs were empowered to deal with all sorts of petty cases according to the tribal custom. In the case of criminal offences of serious nature like violence, death and serious accident, he was to report to the Superintendent.[476] It should

---

[473] *Ibid.*

[474] Imdad Hussain, "Resistance, Pacification and Exclusion; The Hill people and the Nationalist Upsurge", in Arun Bhuyan (Ed), *Nationalist Upsurge in Assam* (Guwahati , 2000), p.282. He further added, "The entire social life of the Lushais was governed by their ideal of 'Tlawmngaihna' by which every individual knew and discharge his responsibilities, including service to the chiefs. In short, no chiefs could make any demand and no Lushai was obliged to render any service beyond what the custom sanctioned."

[475] N Chatterjee, *op.cit.*, p.19.

[476] *Ibid.*, p. 5.

be pointed out however that the Superintendent did not entertain all cases of appeal in accordance with the general principle of non-interference.[477] It was clearly envisaged that the chiefs would function and enjoy powers within limits imposed by the British. Certain traditional rights not compatible with this was abolished by the British government, viz:

1. Right to order capital punishment.
2. Right to seize the property of a villager who was found disloyal or disobedient or one who wished to transfer his allegiance to other chief's *Ram* (territory).
3. Property rights over lands now arbitrarily reserved by the government.
4. Right to tax traders doing business within the chiefs' jurisdiction.
5. Right to freedom of making their sons chiefs under their own jurisdiction.[478]

The division of the powers and responsibilities between the chiefs and the Superintendent in practice, virtually empowered the British to interfere in all aspects of local administration.[479]

The Superintendent was to settle all disputes between the chiefs, between villages, and prevent breach of public peace. He was not to adjudicate the general management of the police, but he would be the head of the police as of all other departments. All arrangements regarding the provisioning of the force and all arrangement of transport would be under his responsibility.

---

[477] *Ibid.*

[478] A G McCall, *op.cit.*, p.202.

[479] *Ibid.*, pp. 202-203."There is no room, whatever, for any sense of perfection in the Government's own approach to the problem, which was, is, and will be, difficult, and which demands far greater care than is generally realised, judging from much which has been written on this interesting and vital problem of the administration of the backward people."

The Superintendent was empowered to regulate the succession of chieftainship, appoint new chiefs and guardians of minor chiefs, partition of the existing villages, punishment and deposition of the chiefs and headmen for misconducts.[480] He was competent to administer civil justice and inflict even a death sentence or transportation for life subject to the confirmation of the Chief Commissioner.[481] He was further empowered to direct any person to abstain from certain acts if he considers such action was likely to disturb public peace and tranquillity.[482] The Superintendent may order any person, not being a native, to leave the district if the presence of such persons was a threat to its peace and goodwill.[483]

The imposition of the House Tax and impressed labour was in fact the symbol of the British power. The house tax imposed in the Lushai Hills district, however, was not the same as had been practised in other plain areas. It was a kind of revenue imposed on every family, but largely less valuable compared to their counterpart in other areas. It was indeed a token of acceptance of the British authority within the Lushai Hills. As the British considered the occupation and submission of the hill tribes was of first importance, it became essential feature of their administration.

The general characteristic of the system was that the house tax was collected at the rate of Re.1/- from every house per year and submitted to the Superintendent by the village chiefs.[484] Earlier in 1891, McCabe proposed a uniform rate of

---

480 V V Rao, "The Sociology of Mizo Politics", in S M Dubey (Ed). *North East India: A Sociological Study*, (New Delhi, 1978), p. 217.

481 *Ibid.*

482 *Ibid.*

483 B C Chakravorty, *British Relations With The Hill Tribe of Assam Since 1885, op.cit.*, p.80. "Five Chin families came across the Tiau River into Lushai Hills and settled on a land belonging to Thansuma, a Chief in the Lunglei Sub-Division. They were fined Rs. 20/- each family and were asked to go away after gathering their harvest. As the houses were still found on the site two month after their harvest had been collected, they were burnt down."

484 FEBP. September 1891; Nos. 243-248, Shakespear, dairy for the week ending 2nd January 1891.

taxation throughout the Lushai Hills, a tax of Re.1/- per house per year should be collected, widows and crippled be exempted; and also a liability to supply rice at the rate of Rs.2/- per maund.[485] His suggestion was accepted by the Commissioner on condition that the Officer in charge was to accept the tax in cash as well as in kind. '*Dhan*' or unhusked rice in place of cleaned rice was also to be accepted at the rate of 20 seers as against ten seers of clean rice.[486]

There was a tendency to increase the house tax from Re.1/- to Rs. 2/- right from its formal imposition. However, the actual decision could not be attained until the local Officers conference of 1896 when a proper rate was finalised and accepted for the assessment of the House tax.[487] As such the house tax was collected accordingly throughout the Lushai Hills district, though exemption of certain persons were contemplated.

The impressed labour was the next important issue. After the submission of all the Lushai chiefs, the British officers exacted impressed labour from the people through their chiefs. It was the only other "general tax" the people owed to the government. Every chief and headmen was to supply labour both paid for and free, on requisition by the Superintendent or his assistants. Both paid labour and the free labour were supplied whenever necessary for a fixed number of days per year and the actual limit depended on the circumstances of the locality.[488]

Even before its proper imposition a number of the officers proposed to lessen the rate of the paid labour and the question of coolie and impressed labour was discussed quite a number

485 FEAP January 1892; Nos. 51-61, McCabe's Eastern Lushai Promenade.

486 MSR. CB 101, Gen. 150, No 23 Letter from the Commissioner of Chittagong to the Chief Secretary to the Governor of Bengal, No. 133 HT of 28 February 1891.

487 *Ibid*., Proceedings of the Chin-Lushai Conference, 1896.

488 MSR. CB 101, Gen. 150 *Op.cit*., McCabe to the Chief Commissioner of Assam.

of times. Expressing his view, John Shakespear had said:

> I think we should insist on each chief supplying a certain number of coolies each year, but I think that we ought to pay these men except when the labour is exacted as punishment. Our object is to encourage the people to work for wages and we shall not do so by making them work for nothing. That a chief supplies the number of coolies asked for is sufficient token of submission and we can afford the men without any loss of prestige. I think under ordinary circumstances we should fix the rate of wages at 4 annas a day.[489]

W B Oldham, Commissioner of Chittagong had proposed the rate of the wages of the labour should be 8 annas a day instead of 4 annas.[490] Nevertheless, A W Davis, Political Officer of the North Lushai Hills, suggested that the rate should be 4 annas a day, which was the average rate with that of the other districts like Naga Hills, he further stated that fixing higher rate than this would only serve to increase difficulties.[491]

It was in 1896 a clear cut proposal for the rate of wages as well as the number of days to be insisted upon was worked out. The Lieutenant Governor of Bengal, after consulting the Chief Commissioner of Assam, enhanced the rate of labour as 8 annas a day from 1897, and to meet this requirement the rate of house tax was also increased from Re. 1/- to Rs. 2/- per house per year.[492]

Except certain persons, viz., government servants, physically handicapped, school masters, chiefs and their officials such as *Upas*, blacksmiths, priests and writers were required to make themselves available, unless they were of old age. It therefore became a burden for the Lushais.

---

[489] MSR. Administrative report of the South Lushai Hills for the year 1891-92.

[490] *Ibid.*

[491] *Ibid.* Administrative report of the North Lushai Hills for the year 1892-93.

[492] D C (A)'s Col-III, No.3 of 1897; Chief Commissioner of Assam Order No. 92 FOR/874 P of 15 March 1897.

Therefore, those exempted from such labours were of highly regarded.

Hitherto, land ownership was traditionally the exclusive right of the chiefs until the British occupation of the Lushai Hills. After the British occupied the hills land became the property of the government and it was leased out to the chiefs.[493] The lease could be withdrawn from the chiefs as a measure of punishment whenever they violated or in any way disobeyed the government orders.[494] A new system of land distribution was introduced in 1898. By this, each chief was given a certain area within which he and his people could move about freely.[495] In addition he was given two or more adjacent village sites and well-defined natural boundaries in order to ascertain his jurisdiction where he would fear no encroachment. Considering the need to fix boundaries of the land of each chief, Shakespear gave them their own "*Ramrilehkha*" or boundary paper.[496]

This land settlement system was perhaps the most important administrative measure introduced by Shakespear. Chiefs now felt secure. The system had many advantages: it removed all causes of dispute among the chiefs as each of them was aware and understood his jurisdiction; the lands being given for life the chiefs tried to get the most out of it and preserved even the forest. Finally the chiefs found that it was to their interest to carefully maintain their inter-village roads.

However, it needs to be pointed out that in some case lands were given to persons other than chief because of their service to the government. This practice did cause a dent in the prestige of the traditional chiefs. But the fact that chiefships

---

[493] Lalrimawia, *op.cit.*, p.104.

[494] Sangkima, *Mizo: Society and Social Change, 1890-1947*, (Guwahati, 1992), p. 118.

[495] B C Chakravorty, *op.cit.*, p.80.

[496] N E Parry, *A Monograph on Lushai Custom and Ceremonies*, (Calcutta, 1927), Reprint-1976, p. 3.

were made hereditary, subject to their good behaviour, and physical and mental capability, ensured their position. The eldest son of the chief was made the rightful heir to his father, again subject to the approval of the Superintendent.[497] Even when new villages were established by the chiefs for their trusted sons it was subject to the approval of the Superintendent. Efforts were made to ensure that the chiefs were trained to administer their subjects in accordance with the standards demanded by the British rule.[498]

Another of Shakespear's administrative measures was division of the two Sub-Divisions into Circles, 12 in Aizawl Sub-division and 6 in Lunglei Sub-division, under the charge of the Circle Interpreter and *Chaprasis*.[499] They were made the channel between the Sub-Divisional Officers, the chiefs and their people.

LUSHAI HILLS MILITARY POLICE

The Superintendent's coercive powers were more visibly displayed in the organisation and distribution of his Military Police. Before annexation of the Lushai Hills Cachar's southern frontier was defended by the Surma Valley Military Police distributed in a chain of frontier outposts. The force, it may be mentioned, grew out of the Frontier Police that came into existence after the reorganisation of the police in India in 1860-61. Cachar Frontier Police was reorganised in 1883-84 and became in 1890 the Surma Valley Military Police. It will be recalled that some men of the battalion moved up to Lushai Hills in 1889 under Cachar's Superintendent of Police W W Daly and established the Aizawl post. After the Chin-Lushai expedition the force was gradually moved up and when the North Lushai Hills came under Assam it became the garrison.

---

497 Parry, *Lushai Custom*, *op.cit.* p. 4.

498 McCall, *op.cit.*, p.244.

499 Robert Reid, *op.cit.* , p.44.

The established policy was to use the military for offensive operations and once pacification was achieved to entrust the responsibility of garrisoning to the Military Police. McCabe's operations against the eastern Lushai during 1891-93 clearly established the usefulness of Military Police composed largely of Gurkhas, for offensive operations. With the result that the military was gradually withdrawn and the defence and security of the North Lushai Hills devolved upon the Military Police.

A special body of Bengal armed Civil Police formed in 1860 to protect and control the Chittagong Hill Tracts, later became the Chittagong Hill Tract Frontier Police, was styled as the South Lushai Hills Military Police Battalion in 1891 with Headquarters at Lunglei.[500] The garrisons were posted at Fort Tregear, Fort Lalthuama, Fort Lunglei, Lungsen and Demagiri with 539 of all ranks. But in 1894 the strength was reduced to 350.[501] The year 1896 witnessed a mutiny in the history of the Military Police in South Lushai Hills owing to grievances among the ranks against the cutting of half of their allowances.[502]

With the amalgamation of North and South Lushai hills under Assam, the South Lushai Hills Military Police Battalion was also amalgamated with the North and came to be known as the Lushai Hills Military Police Battallion.[503] The total strength of the force was however reduced to 970.[504] With the settlement of the Lushai Hills a large force in Lushai Hills was no longer necessary. Therefore, the Military Police were reduced by 150 to a strength of 840 and were armed only with Martini Rifles.[505]

---

[500] *Ibid.*

[501] FEAP. December 1896; Nos. 67-73. Francis H Skrine to the Chief Secretary to the Government of Bengal. 1 September 1896.

[502] *Ibid.*, For details of the Mutiny see Appendix No. V.

[503] L W Shakespear, *op.cit.* Appendix E. p.290.

[504] *Ibid.*, p.243.

[505] *Ibid.*

EPILOGUE

British conquest of the Lushai Hills, as this study shows, was neither easy nor peaceful. During the years from 1890 when the hills were first penetrated to 1896 when they were finally pacified, the Lushais under their chiefs took up arms to resist the imposition of colonial rule. It began with the Western Lushais under Kalkhama and Lianphunga which was followed by that of the Eastern Lushais under Lalburha and Kairuma. In the southern region the uprising was spearheaded by the Southern Sailo chiefs, popularly known as the Haulawngs, and the Pawis or Fanai chiefs, often referred to as the Muallianpuis.

A close examination of the available evidence suggest that Lushai resistance was essentially a refusal on their part to accept the ways in which their political and social life was sought to be defined by the British. Captain Herbert Browne's removal of the chiefs for their complicity in raids in the frontier villages, his imposition of tax and his demands for labour, all of which was foreign to their system, was thus at the root of the disturbances in the western hills. Robert McCabe, the subsequent Political Officer, who suppressed the rising, again extended the same demands on the Eastern Lushai people, which were promptly resisted by them under the leadership of the chief Lalburha. The evidence that Haulawng group would obviously resist the imposition of British colonial rule was revealed when they attempted to block Captain John Shakespear's forward movement to relieve Robert McCabe against the Eastern Lushais at Lalburha's investment. This was in sympathy with their fellow clans in the eastern Lushai hills.

The descendants of Vuta, Sailo chief, more popularly known as Kairuma group, in concert with the Pawis or Fanai chiefs Zaduna and his associates, better known to the British as "Muallianpuis," proved equally troublesome against the British process of pacification of the hills. They refused to supply labour to work under the British to the distant places other than their locality.

Lushai opposition and the leadership of the resistance must be seen in the light of the relationship between the chiefs and the people, which came in conflict with the British policy. The Lushai code of conduct known as *Tlawmngaihna* had been the guiding principle of the community from time immemorial. It provided the need based free and voluntary service within the society without any reward in return. Whereas British colonial rule brought about demands through the chiefs of impressed labour and house tax upon the Lushai people. The Lushai chiefs were not autocratic or despotic rulers. The chiefs exercised enormous powers only during the wars and conflicts but in normal situations there was no sign of such absolute authority. He neither had the power to impose labour nor insist upon their paying house tax. When the British insisted upon all these demands through the chiefs, the Lushais under their chiefs rose up in arms against a system which was quite contrary to their social and political institutions. They totally refused to serve as labour and provide supplies to the British. The chiefs also could not agree to their men serving as labour or providing supplies which would have undermined their authority, nor could they impose any exaction upon their people on behalf of the British for which society gave them no power. When the permanency of colonial rule appeared evident, the chiefs and the people combined to resist the British.

The intensity and protracted nature of the resistance to the British was largely due to the transformation of Lushai warfare by the last decade of the nineteenth century. The expeditions against the Lushais, particularly the Lushai expedition of 1872, had clearly shown the effectiveness of firearms. From the seventies onwards there was an arms traffic in the Lushai Hills and by the time, British decided to annex the Lushai Hills and impose their demands, the use of firearms had become widespread. The change in warfare can also be seen in the exceptionally well-contrived fortifications that withstood even artillery fire.

Not surprisingly therefore, A W Davis, the Officiating Political Officer of the North Lushai Hills, insisted upon the surrender of firearms. The methods employed by the British in suppression of the rising, however, differ from place to place. The Lushais of Eastern, Western, and Haulawng group risings were suppressed through the use of military. The Kairuma group was subjugated mainly by disarmament and the destruction of their villages, livestock and granaries. Eventually, military operations, disarmament, and the political activities on the whole brought about the final pacification of the whole of the Lushai Hills.

The resistance of the different group of Lushais were put down and the hills pacified, however, not before a great deal of effort for disarming the Lushai people. The disarmament tasks would have been difficult had not Bertram S Carey carried out a successful disarmament operations in the Chin Hills. The Lushais, having the Chins as their constant enemy, were not at all ready to surrender their arms until their counterparts in the east were fully disarmed. Disarmament of the Lushai people became the most striking feature of the British process of consolidation. From the beginning of the operations the British took special care to disarm the people. Surrender of guns was at first insisted upon as a punitive fine but soon became an integral part of the pacification process. After the peace was restored and the Lushai Hills District was established the Arms Act was extended to the hills.

In suppressing the uprising the British resorted to wanton violence such as burning of villages and destruction of grain. During the entire operations the British destroyed as many as twenty five villages including the burning of not less than five thousand maunds of grain. McCabe who, in spite of the extensive achievements in his operations against eastern and western Lushais, had expressed that he was not in a position to be proud of his success in stand up fight with the Lushai guerrilla warfare. It was only the destruction of their food supplies and by preventing them from *jhuming* works that the force could enforce an unconditional surrender. The success of the imposition of colonial rule, was in fact, due to the British methods that struck at the very root of Lushai economic life.

While the process of pacification was going on an attempt was made by the authorities in Calcutta to consolidate the hills. The Chin-Lushai Conference raised the question of putting the whole Chin-Lushai country under one administration, but proved to be failure due to certain objection made by the Chief Commissioner of Burma and his officials from the Chin Hills. Resuming the same question the local officers finally decided to amalgamate the North and South Lushai Hills, which took effect on 1 April 1898. The Chief Commissioner of Assam made the proclamation which appointed Captain Shakespear as the first Superintendent of the amalgamated district.

Despite the fact that the Lushai chiefs and their men deliberately resisted the imposition of British colonial rule, at the close of nineties the whole Lushai Hills was consolidated under the British "Indirect rule." The indirect administration introduced by the British in the Lushai hills was not a new system. As early as 1826, David Scott, the Agent to the

Governer General in Assam, proposed the adoption of the indigenous system of village administration. The principle which he tried to follow was "to make as little alteration in existing institution as possible in the eastern part of the country."[506] His main aim was to maintain local institutions and to use local personnel as far as possible. Since then 'non-intervention' was the settled policy of the British towards the hill tribes. In the Naga Hills, the village headmen or the leader of the clans, *Gaoburah* were made responsible in maintaining law and order and to collect the revenue.[507] In the Khasi Hills the *Syiems* were recognised and given his powers to rule over his people. But the British Government had the right to remove a *Syiem* and appoint another in his place on the ground of misgovernment or oppression. At the same time in Garo Hills *Laskars* were made *Zamindars* and empowered to try civil and criminal justice for maintenance of law and order in their respective jurisdiction and to induce the Garos to pay their taxes and tribute.[508]

The intensity of the Lushai resistance clearly showed the importance of the chiefs, and reinforced the concept of rule through chiefs. After the consolidation of the Lushai hills and with the introduction of the "indirect rule" the people were soon isolated from the declining power of the chiefs. The effect of the British occupation of the Lushai hills and influence upon the people were clearly reflected in their songs and ballads. The traditional tune of the songs was fast declining and all the composition of the songs tended to match up with the western style. The life style and dresses also gradually changed with the introduction of the western practices and education.

---

[506] A C Banerjee, "The New Regime", in H.K.Barpujari (Ed.) *The Comprehensive History of Assam*, *Op.cit.* p.28.

[507] J B Bhattacherjee, "Forward Policy", in H K Barpujari *Ibid.*, p. 169.

[508] *Ibid.* p. 170.

APPENDIX-I

Geneological Table of Lushai Chiefs

- Lallula
  - Lalpuiliana
  - Lallianvunga
  - Mangpawrha
  - Vuta

Western Lushai Chiefs
(Kalkhama group)

- Mangpawrha
  - Suakpuilala
    - Lalchhunga
    - Kalkhama
    - Lianphunga
    - Sailianpuia
    - Thanruma
    - Lalhrima
    - Lalkranga
    - Hrangkhupa
      - Zahrawka
      - Dokhama
      - Tulera
      - Lalsavuta
  - Mangpawrha
  - Runphunga

APPENDIX-II

Geneological table of Lushai chiefs.

- Lallula
  - Lalpuiliana
  - Lallianvunga
  - Mangpawrha
  - Vuta

Eastern Lushai Chiefs

(Lalburha Group)

- Lalpuiliana
  - Lalsavunga
    - Vanhnuailiana
      - Dothiauva
      - Liankhama
      - Lalburha
      - Buangtheuva
      - Chinhleia
    - Lalphunga
    - Thangphunga

- Lalphunga
  - Pawibawia
    - Vanphunga
    - Lalruaia
    - Thangkama
  - Lalruma
  - Chunglena

- Thangphunga
  - Thandanga
  - Lalhanga

APPENDIX-III

Geneological table of Lushai chiefs.

Lallula

Lalpuiliana Lallianvunga Mangpawrha Vuta

Eastern Lushai Chiefs

(Kairuma group)

Vuta

Lalnguauva Lalkhuma Lalthanglula Lalhranga Kairuma Lungliana

Ralthianga Lalbuta Nikhama

Lianhminga Zataia

APPENDIX-IV

The Mutiny of the Military Police which occurred on 31 July 1896 was not of without reasons. Robert Henry Sneyd Hutchinson of the Bengal Police took over the Charge of both Superintendent of the South Lushai Hills and Commandant, during the absence of Captain John Shakespear. Skrine and Captain G.H. Loch in their joint letter to the Chief Secretary of Bengal on 1 September 1896 stated thus: "It has been the practice for many years past to employ such sepoys of the battalion as are not detailed for guard, on fatigue duties. Prominent among the latter is that of bringing in bamboos, for which there is a large and constant demand for building purposes. The rate at which they were paid for this service was originally two annas per diem. Abuses appear to have crept in owing to the absence of a fixed tale of work; and the late Commandant, Major Shakespear directed that sepoys employed in fetching bamboos should make two trips per diem, and bring ten before earning their working pay. The latter was subsequently increased to three annas per three bamboos, and since Sneyd Hutchinson's assumption of command the men have been sent out once a day only on this duty. Thus they now get three annas four pies, as compared with two annas under the former system, for bringing in five bamboos; and are saved one daily climb to their quarters from the spurs on which the bamboos are found. But unfortunately no minimum length and girth were prescribed in the bamboos to be paid for; and this omission, as we shall see, led to friction and discontent." The organised Police therefore started their disobedience, "On the 31 July Mr. Williamson ordered that the men should parade in the morning, take their meal and proceed on fatigue duty instead of going out before feeding. His intension was to lessen their labour in view of the greater distance to be traversed by them. Thus at 8:15 am., after parade was over, the men took their meal and went out for bamboos. Hutchinson, the Commandant, returned from Demagiri at 8:30 am. On that day, and at 5 pm. The day officer, Jamadar Gorey Gurung, reported that on the men's return from fatigue duty sixty one of them had left their rifles and belts in the office, declaring that they would do no more work. At the same time a great clamour arouse from the parade ground. Hutchinson put on his uniform and proceeded thither with Williamson and the Subadar Major. He found an exited crowd of sepoys brandishing sticks and shouting that they would beat the Overseer who had reported their work as bad. Seeing that it was useless to expostulate with them in their heated temper, he said that he would hear their grievances on the following morning, warned them that they had broken the law, and retired to his quarters.

Their grievances as stated by them were: 1. They had been fined, on 24 July, half the allowances they had earned for fetching bamboos, though no better were to be got. 2. Subadar Major Bikram Gurng harassed them and treated them unfairly, concealing their complaints. They demanded that he should be instantly removed. 3. Mr. Williamson was too strict with them; they insisted on his transfer. They said that in the event of these demands not being complied with, they would all claim their discharge.

On hearing the incident the Commissioner of Chittagong immediately proceed to South Lushai Hills, and Captain Loch, Commandant of the North Lushai Hills Military Police was sent to control the situation. As soon as he arrived Captain Loch took the command and peace was restored within seven days on 20 August 1896.

GLOSSARY

Aijal	—	Aizawl.
Chaprasis	—	Interpreter.
Dao	—	A two-edged knife about three inches wide at the end and one inch wide near the handle.
Darbar	—	Assembly of chiefs.
Dhaleswari	—	Tlawng river.
Gutoor	—	Tut river.
Fathang	—	Rice tax due to the chiefs.
Fort Aijal	—	Fort Aizawl.
Fort Lungleh	—	Fort Lunglei.
Gyal	—	A species of cow peculiar to the hills.
Hats	—	Markets
Howlong	—	Sailo clan settled in the southern Lushai Hills.
Jahoota	—	Zahuata.
Jhume	—	Shifting cultivation practised in the hilly areas.
Lakhers	—	A Lushai tribe popularly known to the British as Shendus.
Lal	—	Chief.
Lalsiva	—	Lalchheuva.
Lengpunga	—	Lianphunga.
Lungleh	—	Lunglei.
Mantri	—	Minister.
Metnas	—	A species of cow peculiar to the hills.
Mithun	—	Same as above.
Muallianpui	—	Pawis settled around Muallianpui hills.
Muntree	—	Same as mantri.
Naick	—	Lance Corporal.

Ohipum — Uiphum.
Karbari — Same mantri.
Karnafuli — Khawthlangtuipui river.
Khuaichhiah — Bee tax due to the chiefs.
Kolodyne — Chhimtuipui river.
Kooki — Kuki/ Lushai.
Paitu — Paite.
Pem/Pemming — Shifting allegiance to other chief.
Poi — Pawi.
Puithiam — Priest.
Punji — Village.
Ramhual — Chief's adviser as to where the jhums. should be cut.
Rani — Women Chief, or wife of a Chief
Rengti — Rengte.
Sachhiah — Meat tax due to the chiefs.
Saheb — Officer.
Sirthay — Sirte
Sonai — Tuirial river.
Sungoo — Sangau.
Sunnuds — a written authority.
Sylo — Sailo clan.
Tashons — Tlaisun.
Thirdeng — Blacksmith.
Tipai — Tuivai river.
Tlangau — Village-crier.
Upa — Elder or in other words Minister.
Val Upa — Elder among the youth.
Wenon — Hualngo.
Yahow — Zahau.
Zarock — Zahrawka.
Zu — Rice Beer.

BIBLIOGRAPHY

A. PRIMARY SOURCES:

i) Unpublished

National Archives of India, New Delhi

Foreign Department, Political 'A' Proceedings
Foreign Department, Political 'B' Proceedings
Foreign Department, External 'A' Proceedings
Foreign Department, External 'B' Proceedings
Foreign Department, Secret-E Proceedings

Assam Secretariat Records, Guwahati

Foreign Department Proceedings
Home Department Proceedings
Education Department Proceedings
Political Department Proceedings

Mizoram Secretariat Records, Aizawl

Select Papers only

Deputy Commisioner's Records, Aizawl

Select Papers only

Deputy Commisioner's Records, Lunglei

Select Papers only

ii) Published

Accounts, Memoirs and Reports.

Browne, H R : *The Lushais, 1873-1889*, Shillong, 1889.

Carey, Bertram *et.al.* : *The Chin Hills*, Falam,1932;

Dubey, S M (Ed): *North East India: A Sociological Study*, New Delhi,1978.

Elly, E B: *Military Report of The Chin-Lushai Country*, Simla, 1893, Reprint, 1978.

Hunter, W W: *A Life of the Earl of Lord Mayo,* Vol.I, London, 1876.

Hutchinson, R H Sneyd: *Chittagong Hill Tract*, Delhi, 1909, Reprint in 1978.

Lewin, T H: *A Fly on the Wheel (or How I help to govern India)*, Calcutta,1869.

____: *The Hill Tract of Chittagong and The Dwellers Therein*, Calcutta,1869.

Mackenzie, Alexander: *History of the Relation of Government with the Hill Tribes of North East India*, Calcutta, 1884.

Reid, A S: *Chin Lushai Land,* Calcutta, 1893

Shakespear, J: *The Lushei Kuki Clan*, London, 1912.

Soppit, C A : *A Short Account of the Kuki Lushai Tribe,* Bhamo, Burma , 1887 Reprint Cacutta 1976.

Woodthorpe, R.G.: *The Lushai Expedition, 1871-72*, London, 1873.

B. SECONDARY SOURCES

Barkataki, S: *Tribes of Assam*, New Delhi, 1969.

Barpujari, HK : *Problem of Hill Tribes: North East Frontier,* Vols. II & III, Gauhati, 1976 and 1981.

____: *The Comprehensive History of Assam,* Vol. IV, Guwahati, 1992.

Barpujari, SK: "Mizoram Freedom Fighter, Kalkhama And Lianphunga," Paper Presented in the International Seminar On "*Studies On The Minority Nationalities of North East India, The Mizos,*" 1992.

Bawitlung, Vanlalringa: "Origin and Development of Chieftainship in Mizoram," Unpublished, Ph. D. Thesis, North Eastern Hill University, Shillong , 1991.

Baveja, J D : *The Land Where Bamboo Flowers*, Gauhati , 1970.

Bhuyan, Arun, (Ed): *Nationalist Upsurge in Assam*, Guwahati, 2000.

Buckland, CE: *Bengal Under Lieutenant Governor,* Vol. I. Calcutta , 1901.

Buckland, C E : *Dictionary of Indian Biography*, London, 1905, Reprint, New Delhi, 1971.

Chakravorty, BC: *British Relation with the Hill Tribe of Assam*, Calcutta, 1958.

Chandra, Bipan: *Essays on Colonialism*, New Delhi, 1999.

Chapman and Clark: *Mizo Miracle*, Madras, 1968.

Chatterjee, Suhas: *Mizoram Under British Rule*, Delhi, 1985.

—, N.—: *The Mizo Chiefs and his Administration*, Aizawl, 1975.

—,—: *Zawlbuk as a Social Institution in the Mizo Society*, Aizawl, 1992.

Chaube, SK: *Hill Politics in North East India*, Orient Longman, 1973.

Frontiers and Overseas Expedition From India, Vol. VII, The Official Account of the Abor Expedition, (Reprint) Delhi, 1983.

Gailey, Harry A: *History of Africa from 1800 to Present*, New York, 1972.

Goswami, B B : *The Mizo Unrest*, Jaipur, 1979.

—,—: "Social and Political Institutions of the Mizo." Proceedings of a Seminar on *Social and Political Institutions of the Hill People of North East India*, Organised by NEICSSR, Shillong and Sponsored in Collaboration With Anthropological Survey of India, Calcutta. Published by Anth. Survey of India, 1977.

Hluna, JV: *Church and Political Upheaval in Mizoram*, Aizawl, 1985.

Kyndiah, P R: *Mizo Freedom Fighter*, New Delhi, 1994.

Lalchungnunga: *Mizoram : Politics of Regionalism and National Integration*, New Delhi, 1994.

Lal D dena: *Christian Mission and Colonialism: A Study of the Missionary movement in North East India with particular reference to Manipur and Lushai Hills 1894-1947*, Shillong, 1988.

Lalremsiama, F: "The Traditional Political Institute of the Lusheis," Unpublished M.Phil. Dissertation, North East Hill University, Shillong,1984.

Lalrimawia,: *Mizoram: History and Cultural Identity*, Guwahati,1995.

—,—: "Lushai Hills Administrative Development, 1894-1947," Unpublished Ph.D. Thesis, Gauhati University, Gauhati , 1980.

Lalsangpuia, H S : "Frontier Policy under Lord Mayo with special reference to the Lushai in the North East Frontier 1868-1872," Unpublished M.Phil. Dissertation North East Hill University, Shillong, 1983.

Lehman, FK: *The Structure of Chin Society*, University of Illinois, 1963, Reprint in 1980.
Lianzela: *Economic Development of Mizoram*, New Delhi, 1994.
Lorrain, James Herbert.: *Dictionary of the Lushai Language,* Calcutta, 1940.
—,— Reginal A : *5 Years in Unknown Jungle*, London, 1912.
Low, DA : *Lion Rampant*, London, 1973.
Luthra, PN: *Constitutional and Administrative Growth of the North East Frontier Agency*, Shillong, 1971.
Macrae, John: "Account of the Kookies or Lunctas," *Asiatic Researches*. Vol. VII. No. VI; Calcutta, 1801,
McCall, Major A G : *Lushai Chrysalis*, Shillong,1949.
Nag, Chitta Ranjan: *The Mizo Society in Transition*, New Delhi, 1993.
Nunthara, C : *Mizoram: Society and Polity*, New Delhi, 1996.
Pachuau, Rintluanga,: *Geography of Mizoram*, Aizawl, 1994.
Parry, N E: *A Monograph on Lushai Custom and Ceremonies*, Calcutta, 1928. Reprint, 1976
—,—: *The Lakhers*, Shillong, 1932, Reprint, 1988.
Prasad, Dr RN: "Traditional Political Institution of Chieftainship in Mizoram: Power, Function and Privileges," Paper presented in the International Seminar on *Studies on the Minority Nationalities of North East India*, 1992.
Prasad, RN: *Government and Politics in Mizoram*, New Delhi, 1987.
Dubey, SM (Ed): *North East India: A Sociological Study*, New Delhi, 1978.
Ray, Animesh: *Mizoram: Dynamics of Change*, Calcutta, 1982.
—,—: *Mizoram*, Calcutta, 1993.
Reid, Sir Robert: *History of the Frontier Areas Bordering on Assam,1883-1941*, Shillong, 1942.
Rigby, Captain G C : *History of Operation in Northern Arakan and the Yawdwin Chin Hills, 1896-97*, Government Printing Press, Burma, 1897, Reprint, 2000.
Rawlins, John: "On the Manners, Religion and Laws of the Cucis or Mountaineers of Tipra," *Asiatic Researches*, Vol. II, No. XII. Calcutta, 1972.
Sangkima: *Mizos: Society and Social Change*,Guwahati, 1992.
Sen, Sipra: *Tribes of Mizoram, Description,Ethnology and Bibliography, 1840-1990*, New Delhi, 1992.

Shakespear, LW: *History of the Assam Rifles*, London, 1929.
Singh, Bhupinder: *Antiquity to Modernity in Tribal India*, Vol. II. (Tribal Self-Management in North-East India), New Delhi, 1998.
—,—Daman: *The Last Frontier: People and Forests in Mizoram*, New Delhi, 1996.
Singh, Jai Prakash, *et.al.*: *Archeology of North Eastern India*, New Delhi, 1991.
—,— SN: *Mizoram (Historical, Geographical, Social Economic, Political and Administrative)*, New Delhi, 1994.
Stevenson, H N C: *The Economics of the Central Chin Tribes*, Falam, 1943, Reprint, 1986.
Syiemlieh, DR: *British Administration in Meghalaya: Policy and Pattern*, New Delhi,1989.
Thanga, LB: *The Mizos: A Studies in Racial Personalities*, Gauhati,1978.
Tribal Research Institute, Government of Mizoram: *Know Your Own Land, Vols. I-IV*, Aizawl, 1982.
—,—: *Paite in Mizoram*, Aizawl, 1987.
—,—: *Foreign and Political Department Report 1874*, Aizawl, 1980.
—,—: *Foreign Department Report, 1892,* Aizawl, 1980.
—,—: *Monoliths and Landmarks of Mizoram*. Aizawl, 1979.
Vanlalauvi: "The Traditional Mizo Society," Unpublished M.Phil. Dissertation, Department of History, North Eastern Hill University, Shillong, 1988.
Verghese, C G *etal. A History of the Mizos*, Vol. I, New Delhi, 1997.
Vumson : *Zo History*, Aizawl, 1986.

IN MIZO

Bawichhuaka, H K : *Mizo Ziarang*, Mizo Academy of Letters, 1981.
Challiana, Rev : *Pi Pu Nun*, Aizawl, 1969.
Chaltuahkhuma : *Political History of Mizoram*, Aizawl, 1981.
Chawngkunga, C : *Genealogical Tree of Mizoram*, Aizawl, 1996.
Dokhuma, James : *Zokhaw Nun*, Aizawl, 1998.
Doliana, Dr. K L : *Mizo Nunhlui leh Hlate*, Aizawl, 1988.

Hualngo Literature and Cultural Association: *Zofate Chanchin*, Tahan-Kalemyo, 1995.
Lalhmuaka: *Zoram History*, Aizawl, 1992.
Lalremsiama, F: *Milu Lak leh Vai Run*, Aizawl, 1997.
Lalrinawma, H: *Hmasang Zofate Chanchin*, Aizawl, 2000.
Lalsangpuii: *Hmanlai Mizo Awmdan* (Pu Thanga Ziak), Aizawl,1992.
Lalthangliana, B: *Hmasang Zonun*, Aizawl, 1992.
Lalthanliana: *Mizo Chanchin*, Aizawl, 2000.
Liangkhaia, Rev: *Mizo Chanchin*, Aizawl, 1938.
Lunghnema, Col. V: *Mizo Chronicle*, Churachandpur, 1993
Rokhuma, K.L: *Mizoram Zirnaa Mission leh Kohhran Rawngbawlna*, Aizawl, 2000.
Sailo, Lalsangzuali: *Lalnu Ropuiliani*, Aizawl, 2000.
Saiaithanga: *Mizo Kohhran Chanchin*, Aizawl, 1969.
Siama, V L: *Mizo History*, Aizaw, 1953.
Thanglura, A: *Mizoram Politics Inlumleh Dan*, Aizawl, 1989.
Tribal Research Institute, Government of Mizoram,: *Pawi Chanchin*, Aizawl, 1988.
—,—: *Mizo Lal Ropuite*, Aizawl, 1996
—,—: *Mizo-te Khawsak Phung*, Aizawl, 1993.
—,—: *Mizo Ram Chhuah Dan*, Aizawl, 1993.
Vanchhunga: *Lusei leh A vela Hnamdangte Chanchin*, Aizawl, 1955.
Vanlawma, R: *Ka Ram Leh Kei (My Country and I)*, Aizawl,1972.
Vanthuama, H: *Mizoram Politics Chanchin, (1952 Hmalam)*, Aizawl, 2001.
Zama, LT: *Zoram Hun Laita (1930-1945)*, Aizawl, 1991.
Zatluanga: *Mizo Chanchin*, Aizawl, 1966.
Zawla, K: *Mizo Pi Pute leh An Thlahte Chanchin*, Aizawl, 1964.

INDEX